Usability and Internationalization
of Information Technology

Human Factors and Ergonomics
Gavriel Salvendy, Series Editor

Aykin, N. (Ed.): *Usability and Internationalization of Information Technology.*

Hendrick, H., and Kleiner, B. (Eds.): *Macroergonomics: Theory, Methods and Applications.*

Hollnagel, E. (Ed.): *Handbook of Cognitive Task Design.*

Jacko, J.A., and Sears, A. (Eds.): *The Human-Computer Interaction Handbook: Fundamentals, Evolving Technologies and Emerging Applications.*

Meister, D. (Au.): *Conceptual Foundations of Human Factors Measurement.*

Meister, D., and Enderwick, T. (Eds.): *Human Factors in System Design, Development, and Testing.*

Stanney, K. M. (Ed.): *Handbook of Virtual Environments: Design, Implementation, and Applications.*

Stephanidis, C. (Ed.): *User Interfaces for All: Concepts, Methods, and Tools.*

Ye, Nong (Ed.): *The Handbook of Data Mining.*

Also in This Series

HCI INTERNATIONAL 1999 Proceedings 2 Volume Set
- **Bullinger, H.-J., and Ziegler, J. (Eds.):** *Human-Computer Interaction: Ergonomics and User Interfaces.*
- **Bullinger, H.-J., and Ziegler, J. (Eds.):** *Human-Computer Interaction: Communication, Cooperation, and Application Design.*

HCI INTERNATIONAL 2001 Proceedings 3 Volume Set
- **Smith, M.J., Salvendy, G., Harris, D., and Koubek, R.J. (Eds.):** *Usability Evaluation and Interface Design: Cognitive Engineering, Intelligent Agents and Virtual Reality.*
- **Smith, M.J., and Salvendy, G. (Eds.):** *Systems, Social and Internationalization Design Aspects of Human-Computer Interaction.*
- **Stephanidis, C. (Ed.):** *Universal Access in HCI: Towards an Information Society for All.*

HCI INTERNATIONAL 2003 Proceedings 4 Volume Set
- **Jacko, J.A. and Stephanidis, C. (Eds.):** *Human-Computer Interaction: Theory and Practice (Part I).*
- **Stephanidis, C. and Jacko, J.A. (Eds.):** *Human-Computer Interaction: Theory and Practice (Part II).*
- **Harris, D., Duffy, V., Smith, M. and Stephanidis, C. (Eds.):** *Human-Centered Computing: Cognitive, Social and Ergonomic Aspects.*
- **Stephanidis, C. (Ed.):** *Universal Access in HCI: Inclusive Design in the Information Society.*

Usability and Internationalization of Information Technology

Edited by

Nuray Aykin
Siemens Corporate Research, Princeton, NJ

LEA

2005

LAWRENCE ERLBAUM ASSOCIATES, PUBLISHERS
Mahwah, New Jersey   London

Lawrence Erlbaum Associates, Inc., Publishers
10 Industrial Avenue
Mahwah, New Jersey 07430

Cover design by Sean Trane Sciarrone

Library of Congress Cataloging-in-Publication Data

Usability and internationalization of information technology /
   edited by Nuray Aykin.
        p.   cm.
   Includes bibliographical references and index.
   ISBN 0-8058-4478-3 (c. : alk. paper)
   ISBN 0-8058-4479-1 (pbk. : alk. Paper)
      1. User interfaces (Computer systems)   I. Aykin, Nuray.

QA76.9.U83U79 2004
005.4'37—dc22
                                                       2003064338
                                                              CIP

Books published by Lawrence Erlbaum Associates are printed on acid-free paper, and their bindings are chosen for strength and durability.

Printed in the United States of America
10 9 8 7 6 5 4 3 2 1

Contents in Brief

Contents

3. User Interface Design and Culture   51

4. Synthesizing the Literature on Cultural Values   79

Series Foreword

With the rapid introduction of highly sophisticated computers, (tele)communication, service, and manufacturing systems, a major shift has occurred in the way people use technology and work with it. The objective of this book series on Human Factors and Ergonomics is to provide researchers and practitioners a platform where important issues related to these changes can be discussed, and methods and recommendations can be presented for ensuring that emerging technologies provide increased productivity, quality, satisfaction, safety, and health in the new workplace and in the Information Society.

The present volume is published at a very opportune time when industry and commerce is internationalized and products and services are marketed across national and cultural boundaries. This book provides a theoretical foundation and practical examples and guidelines for designing information technology for different cultures, languages, and economic standing and the methods to be utilized to ensure their usability across cultural boundaries.

The book's 11 chapters are authored by some of the foremost practitioners from the United States, Europe, and Asia. The book starts with two chapters on guidelines and practices for internationalization and localization, and includes four chapters on cultural consideration and guidelines in the design of information technology; two chapters on usability evaluation methodology and cost-benefit analysis for cross-cultural design; and three chapters on case studies of cross-cultural design. The book's 368 references, 62 figures, and 45 tables provide useful modes for further in-depth study of the subject.

The book's editor is a leading international figure in the subject area. The book should be of special value to people designing and evaluating information technology products and services for international use and for researchers in the subject area.

—Gavriel Salvendy
Series Editor

xiii

Foreword

You are probably wondering about how to better serve the international users of your product, Web site, or intranet. Why else would you have bought this book? :-) Simply being aware that there is an issue is a great first step. A lot of designers and project managers have no clue that people in other countries are different and need special attention in a user interface design project.

I have three simple messages:

1. International usability is not small potatoes; it has a major impact on your success.
2. Truly serving international users well is a deep design problem that requires serious work.
3. Despite message #2, you should not give up if you only have a small budget to devote to international usability.

First, what's the magnitude of the problem? In one project,[1] we tested 20 American e-commerce sites with both American and European users. The users' ability to successfully shop on the sites was 61% on the average for American users and only 47% for European users. In other words, these sites would be able to increase their overseas sales by almost 33% if only they could deliver the same usability to international customers as to domestic customers. Averaged across several studies we have conducted, measured usability was 46% higher for domestic users than for international users.

Second, why is international usability a deep design problem that relates to more than simply translation? Because a user interface is acted on, not simply read. In one of our projects, we tested the Web site for a big European bank in a variety of countries, including the United States and several European countries. The Web site was available in several languages, in-

---

[1] For more information on this project and the detailed guidelines that resulted from it, please see www.nngroup.com/reports/ecommerce.

cluding English, and the translation was of excellent quality (contrary to the sloppy translation jobs we often see). Still, the American users had great difficulties using the site, because of a variety of cultural differences between the United States and Europe. For example, the American users could not use the feature for choosing their language because it was presented in a way that made sense to Europeans but not to Americans who are used to always seeing everything in a single language. The English version of the site was great in terms of language, but the user interface elements didn't work for American users.

In another example, we tested a customer service site for an American technology product in Korea. Even though the interface was in Korean, the Korean users had difficulty navigating the site because they had difficulty understanding the information architecture (i.e., the way the information was structured).

In a user interface, it's not enough to understand the words. You have to understand the features and you have to be able to navigate the information space. Web sites and intranets are particularly vulnerable to difficulties introduced by poor international usability: One wrong click and you are in the wrong area of the site, but you may never discover that.

The basic tenet of usability is that good design depends on the users and their tasks. People in different countries will often approach similar tasks in different ways because of cultural differences. Furthermore, the users are certainly different. When a user interface is being used in a different country, it's really a different user interface, whether or not it has been translated. With different users and different tasks come different usability requirements for the design. That's why international usability doesn't come simply from wishing for it. Catering to international users requires international usability studies: You need to modify your design to handle those usability problems that have been discovered by testing users in each of your important target countries.

Third, isn't this impossible? I just said to "test users in each important target country," but that would cost a fortune for any company that tries to target more than two or three countries. Yes, truly ideal international usability will indeed be expensive. There are hundreds of countries in the world, and each will have its own considerations. Unfortunately it's also several times as expensive to conduct international user testing as it is to test with local users in your own country.

Despite these facts, don't despair. It is possible to conduct most usability activities domestically, where user research is cheap, and supplement with a smaller number of smaller studies in a few countries, sampling the main regions of the world. At a minimum, make sure that you have test data from the Big Three: United States, Japan, and Germany, all of which are very different. This strategy will not be ideal, and I don't think it will be considered acceptable 20 years from now, but for now it works. You can increase usability substantially for your most important customers in the most important countries through a fairly small effort. They will not get the same quality as your domestic customers, but they will get much better quality than if you hadn't done any international testing. Even customers in smaller countries where you do not test will still benefit somewhat from your increased awareness of the needs of users outside your own country.

Any international usability efforts are better than none. If you only have a small budget for international usability, you may not be able to close the entire 46% quality gap between the domestic user experience and the international one. But if you can just cut the problem in half, you will now be serving your international customers 23% better than you did in the past. That's surely worth a small investment, considering the market potential of the rest of the world.

—*Jakob Nielsen*
www.useit.com

Preface

On entering a country, ask what is forbidden;

on entering a village, ask what are the customs;

on entering a private house, ask what should not be mentioned.

—Chinese Proverb

If this book were published about 10 or even 5 years ago, I would be spending a lot of time explaining why internationalization is so important. And I would need to provide all those statistics about the trends on how companies are going global, how e-business is shaping up global commerce, and how the number of multilingual sites is increasing rapidly. It is not that I do not believe in providing statistics to show the readers the importance of globalization in design. There is no need now to provide all those statistics to convince people about the importance of being part of the global market trends. It is happening now and it is everywhere. Not designing with language and cultural differences in mind is no longer an option that companies can afford. The second reason is that the statistics reflect the moment they are referred to and by the time this book is published, any statistics shown here will be almost obsolete given the fast trends in the Internet industry. However, there is one clear fact that remains the same: Only 8% to 10% of the world's population speaks English as their primary language. Even in the United States, there are Spanish and Japanese TV stations that address the needs of the high percentage of non-English-speaking ethnic groups. In cases where two countries speak the same language, a design for one may not work well in the other. Designers often face questions such as, "My customers read English, so why should I localize (or should we spell it *localise*) my product?" The answer is easy. Just superficially looking at the differences in the language and time/date formatting, we can clearly see issues that would lead to confusion. For example, the United Kingdom uses the 24-hour clock, whereas the United States uses a 12-hour clock; *centre* in the United Kingdom is spelled *center* in U.S. English; and the *toilet* in the United Kingdom means *bathroom* in the United States.

Today, more and more web sites are providing content in multiple languages, and more and more products are being designed with cultural differences in mind. The concept of cross-cultural design, however, has not yet become a priority issue on the practitioners' and educators' agenda. There is a lot of work that still needs to be done to make internationalization a standard part of the design process that lead to successful product launches in different parts of the world.

I remember a full page advertisement that the Japan Automobile Association placed in the *New York Times* in 1995. They listed reasons why American cars were not selling in Japan. The list included a number of bloopers that were not thought through clearly by the U.S. automobile manufacturers: (a) the size of the cars was not practical, (b) the steering wheel was on the wrong side, (c) seat sizes and adjustments did not fit the average Japanese, (d) engines weren't designed to run on Japan's lower octane fuel and thus performed poorly, (e) the letters R, D, and L on the shift had no meaning for Japanese drivers, and (f) repairs required hard-to-find English-based tools.

Everyone has his or her favorite examples related to translations. In some cases, bloopers even occur when a company uses its native language. For example, in 1997, Reebok named their new sneaker line *Incubus,* which means "an evil spirit supposed to descend upon and have sexual intercourse with women as they sleep." When the gaffe was disclosed, they had to trash all packaging and advertising. The moral is: We need to do our homework very carefully, even when we think we know our language well.

Another great blooper is the use of flags to represent languages on Web sites. Flags represent countries, not languages. In some cases, a language is shared by many countries. For example, English, although there are differences, is the language in Great Britain, the United States, Australia, and New Zealand. In other cases, some countries have multiple languages. For example, the official languages in Belgium are Dutch, French, and German. There are also cultural and political issues regarding how a nation treats its flag, and we should be very careful about how we use flags in our designs.

In using examples, however, we need to be careful about not to fall into the urban legends. We see examples of bloopers related to pitfalls in doing business in other countries. One of the most widely told stories is about the Chevy Nova. According to sources, the car did not sell well in Spanish-speaking countries because its name translates as "doesn't go" in Spanish. This, however, is actually an urban legend. The car did sell well in Spanish-speaking countries. "Nova" in Spanish means "star" as in English.

The phrase "doesn't go" is *no va* in Spanish and has nothing to do with the word *Nova*. There is even a government-owned oil company in Mexico, Pemex, that sells gasoline under the name of Nova.

Although many designers and developers do tackle various issues on internationalization, the main questions remain the same: What does it take to design products and services, including software applications and Internet sites, for multiple markets? How much will it cost? Can we justify the cost?

As the Internet has moved toward ubiquity, the need for internationalization is apparent. Initially in the United States, designers felt no imminent need to make sites available in multiple languages. The rationalization was that "the world knows English." Perhaps this is no longer true. Why else would Microsoft spend millions of dollars to release Microsoft Office XP in more than 40 languages?

While several books have been written since the late 1980s on internationalization and localization, this remains an ever-expanding field and there is no book that covers it all. Some emphasize the developers' and software engineers' perspective, some emphasize management's perspective, and some emphasize the designers' perspective.

Our goal, as contributors to this book, is to provide designers with concrete tools and processes—design practices, guidelines, case studies, and lessons learned. Further, we hope to provide you with extensive resources to aid you in your work. This book does not attempt to cover all design aspects of internationalization and localization, and it is definitely not intended to be a software internationalization reference book. We list those books, along with the design-related ones, in the appendix section at the end. As designers, we need to be intimately familiar with the software internationalization terminology and processes, but we should also emphasize the cultural studies that really help us understand cultures in a much more detailed level than just learning to choose the right colors from tables showing color–culture pairs. And it is not just using the right data formatting (e.g., date and time) provided in software tools either. We all believe that this is a vast and wonderful area to explore, and we invite our readers to share their experiences and learnings with others. The more we learn, the more we will make a difference in design practices.

This book is organized into three sections. Part I includes four chapters. Chapter 1, by Aykin, approaches the topic from a management perspective, defining roles and responsibilities. Chapter 2, by Aykin and Milewski, covers the practical issues and guidelines regarding international information displays, what we call the tip of the iceberg of internationalization and localization. Chapters 3, 4, and 5 go deeper into the lower layers of the iceberg and talk about the cultural dimensions and the impact of culture on design. Chapter 3, by Marcus, looks in depth at Hofstede's study on cultural dimensions, while chapter 4, by Gould, describes the studies looking at and theories regarding cultural differences. In chapter 5, Woods describes how to manage multilingual content in the global enterprise setting, as well as methodologies for managing the internationalization and localization of web sites.

Part II focuses on some specific aspects of design, cost justification, and international usability evaluations. In chapter 6, Horton provides a vast array of information on global icon design and includes lots of examples and do's and don'ts that are very useful for designers. Dray and Siegel put all their experiences, along with their knowledge regarding international user studies, into chapter 7. They provide not only a good perspective on conducting international user studies, but also a lot of practical tips for the practitioner. In chapter 8, Mayhew and Bias apply their expertise in cost-justifying usability to cross-cultural user interface design, which costs much more than designing for a single target market.

Part III covers three case studies on international user interface design. In chapter 9, Clarke describes the differences in web site design for children in different countries. Chapter 10, by Röse, looks into various design issues for China. In chapter 11, Degen, Pedell, Lubin, and Zheng discuss designing a travel reservations site for three different countries: Germany, the United States, and China.

At the end of this book is a list of books and magazines on the topic of internationalization in user interface design, project management, and software design and development. Other valuable resources, leading journals, and nonprofit organizations that deal with internationalization are also listed.

I regret not being able to include more topics on internationalization and localization from the design perspective. However, I hope this book will become one of those dog-eared, often used volumes on your bookshelf and remind you that the world is an amazing place where we learn something new every day. As the Dalai Lama's Instructions for Life says: "**Once a year, go someplace you've never been before.**"

And I hope this book will show that the internationalization and localization do not merely mean translation of content, or redesigning and developing an entire new product every time a new market is identified, or just expecting your local offices to handle the adaptation of your product to their markets. It is much more than that, and we hope we have touched many aspects of it.

ACKNOWLEDGMENTS

I would like to thank Gavriel Salvendy for encouraging me to edit a book on this topic and providing in-depth comments on content. My thanks also go to LEA, especially Anne Duffy, Kristin Duch, and Sarah Wahlert for publishing this book and working hard with us to make it happen. I have been grateful for the suggestions and insights I received from my long-time friend, Diane Z. Lehder, who also shared my early years in evangelizing internationalization at AT&T.

Many special thanks go to my husband, Al, and my son Bora for giving me the time and for understanding the sacrifices they made while supporting me in this effort.

PART I

INTRODUCTION
AND CULTURAL CONSIDERATIONS

Chapter 1

Overview: Where to Start and What to Consider

Nuray Aykin
Siemens Corporate Research, Princeton, NJ

HOW AND WHERE TO START?

In this book, the contributors discuss many aspects of internationalization and localization and provide valuable insights regarding cultural differences, design practices, globalization, content management, and so on. This chapter is intended to give the reader an overall firm basis for understanding why we need to localize our products to target cultures, and how to begin doing so. In addition to summarizing key points made by all the authors throughout this book, it also addresses the roles and responsibilities within an organization dealing with internationalization and localization, and provides an action checklist along with practical "how to" guidelines.

The content provided in this chapter is by no means complete. The globalization of a product has many facets and it is almost impossible to cover them all here. However, this chapter, and also this book, is a good starting point when your company decides to move into global markets with the goals of making products suitable, usable, and preferred by the users in the target locale.

WHY INTERNATIONALIZATION AND LOCALIZATION?

The world is a diverse place. The environment in which software applications are designed often is not. It takes a Herculean effort to keep a design team focused on what is required to successfully design products for global markets. But, the change is here, the challenge is real, and we need to face it. The facts about differences in languages and cultures are spread throughout many resources. Consider these facts:

- Only 8% to 10% of the world's population of 6.1 billion speaks English as its primary language.
- In some cities in the United States, large populations speak a language other than English as their native language: 78% in Miami, 49% in Los Angeles, 45% in San Francisco, and 42% in New York City (25% of subway users in New York City do not speak any English at all). Eighty percent of European-based corporate sites are multilingual (Dunlop, 1999). Globally, there are 580 million people with Internet access as compared to 563 million in the last quarter of 2002: In the United States, 29% of the population has Internet access. This compares to 23% in Europe, 13% in Asia-Pacific, and 2% in Latin America (CIO, 2003).
- U.S.-based Palm has a 68% share of the Latin American market (CIO, 2002a).
- The United States is responsible for shipping 38.8% of personal computers worldwide, followed by Europe with 25%, and Asia-Pacific with 12%. The total personal computers shipped worldwide reached 1 billion in 2002 (CIO, 2002b).

Whether to expand to global markets and localize your product for target locales is your company's decision, and it should be based on business needs, user needs, and technology considerations. But if the decision is to go global, your company will need to define new roles and processes if it is to carry out the internationalization process successfully. In this chapter, I touch on these issues. Additionally, I cover some legal issues you need to consider when expanding into global markets, especially in designing global Web sites.

TERMINOLOGY

As defined by the Localization Industry Standards Association (LISA) in *The Localization Industry Primer* (LISA, 2001), here are definitions of the basic terminology used throughout this book.

Globalization: The general process of worldwide economic, political, technological, and social integration. More specifically, the process of making all the necessary technical, managerial, personnel, marketing, and other enterprise decisions necessary to facilitate localization. Globalization addresses the business issues associated with taking a product global. It involves integrating localization throughout a company including product design and development, as well as marketing, sales, and support in the world market.

Internationalization (I18N): The process of ensuring, at a technical/design level, that a product can be easily localized. Internationalization is thus part of globalization. Internationalization helps define the core content and processes so that they can be easily modified for localization to specific markets. To abbreviate the lengthy word *Internationalization,* the term *I18N* was introduced in the early 1990s. The I and N represent the first and last letters of the word *Internationalization,* and 18 represents the number of letters between I and N.

Localization (L10N): The process of modifying products or services to accommodate differences in distinct markets. The term *L10N* was derived in the same way as I18N. Localization makes products or services usable in, and therefore acceptable by, target cultures.

There are two other terms that are related to the field and that are encountered often:

Translation: The linguistic component of localization. It is choosing the appropriate text in a target language to convey the proper meaning from a source language text.

Locale: That part of the user's environment that is dependent on language, country/region, and cultural conventions. For example, Argentina, Spain, and Mexico represent three different locales even though their common language is Spanish.

THE GLOBALIZATION PROCESS

Because globalization involves "the process of making all the necessary technical, managerial, personnel, marketing, and other enterprise decisions necessary to facilitate localization" (LISA, 2001, p. 34), it is important to understand where to start.

The first question a company must address is whether it makes sense to expand into global markets. If the answer is yes, then the company must define global business strategy, which includes identifying target markets, and how to enter them, and restructuring the organization to facilitate globalization efforts. These initial steps are beyond the scope of this book. There are excellent books and references available on this topic; see, for example, O'Hara-Devereaux and Johansen (1994), Rhinesmith (1993), and DePalma (2002). Instead, this book concentrates on the internationalization and localization parts of this process, which include (a) ensuring that your product and web site are designed and developed free of any culture-specific attributes, so they can be easily localized (Internationalization), and (b) making the necessary resource files and design changes so your products and web site are culturally and technically suited for the target culture (Localization).

The internationalization and localization processes follow the same basic steps. First, define the core product or web site that will be localized for the target markets. This involves defining the software architecture, common software, and user interface elements that will constitute the core, regardless of the target locale. The resource files that define the target locale's specific elements, as well as the common global user interface design that accommodates different languages, text expansion, navigation structures, and different content within a corporate defined look and feel (without sacrificing brand) are all considered in this step. If there is no existing product, you can define the core product to be free of any cultural elements. It will then be easier to localize the components for different cultures.

For example, assume date and time will be displayed on a web site. If the software developers use separable resource variables for date and time rather than the fixed U.S. date and time in the code, it will be easier to retrieve the correct date and time representation from the resource files for a particular target locale and place them in the software when needed, even during runtime. If the user interface designers can leave enough space on the screen to accommodate different lengths and formats in displaying date and time, the insertion can be accomplished without changing the layout design. In most cases, however, the products are designed for one country at the beginning, and the need to globalize comes later. In this case, the product must be reengineered, which includes conducting a code inspection to make the software internationalized. For example, if the date and time code is fixed in the core product, it would be replaced by resource variables so that resource files for a particular target locale could be inserted to provide the correct date and time on the screen.

Second, localize the internationalized core product for different target locales. Using the core product and core user interface design guidelines, software designers and developers can add the locale-specific formats and content to the design. In this way, the usage of common elements is optimized. Internationalizing the product first facilitates the localization of the product and makes it more likely that the final product will meet your customers' needs and expectations.

Creating successfully localized products requires a tightly integrated corporate team, and successful integration is dependent on clearly defined roles, responsibilities, and tasks. These are discussed in the remainder of this chapter.

GLOBALIZATION TEAMS: A NEED FOR CHANGE IN THE ORGANIZATION

If your company is moving into global markets and your products need localization, then it is time to start building the corporate infrastructure to carry out the necessary tasks. When a company goes global, all units within the company need to understand the meaning and scope of going global and the impact on their own roles and responsibilities. It is definitely not merely having your product manuals or your web site content translated into a target language, without any changes. While this is a common strategy, it really is not the way to meet the needs and expectations of customers "over there" (a place we sometimes know little about). Starting from the market analysis and continuing until the localized versions of the product are out the door there is a lot of work to do. This work does not end with maintenance and the definition of future releases; it continues as long as we operate in outside markets.

In this section, I talk about the roles in the organization and the changes needed to accommodate the globalization process. The following teams in an organization carry out specific roles in the internationalization and localization processes. Of course, it is possible that some of these teams can be combined under one umbrella. However, the roles and the work that needs to be done should not be sacrificed to save time and money. If it is not done right at the beginning, getting it right later will be far more costly.

- Management Team: Without buy-in from the management team and without a clear understanding of what it takes to internationalize and localize a product, it is very difficult to get the synergy and enthusiasm along with a commitment from the other teams. The management team includes all top-level managers, including country managers if you have a global presence. It is very important to identify a champion who is also accountable for the company's globalization efforts. This champion will ensure the globalization process is overseen properly and will serve as the globalization advocate within the company.
- Marketing Team: The marketing department can easily play the lead role in globalization efforts. They understand and select target countries, define business needs, conduct research/studies in the target cultures to understand needs and preferences, perform branding analyses, validate globalization decisions by defining the return on investment (ROI) from moving into global targets, help the sales team to form distribution channels or partnerships, and work with the legal department to understand the target country's standards, legal issues, customs, and business conduct.

- Implementation (Internationalization/Localization) Team: This is the core team that will work on day-to-day activities as well as the overall project management. While actual team size depends on the complexity and the magnitude of work involved, each implementation team should include:
  - Project managers—who can manage the entire process of internationalization and localization.
  - User interface designers—who can design user interfaces, and know how to study the needs and preferences in the target cultures. Their work requires an understanding of languages spoken, cultural preferences, interests in the types of products and services you offer, use of colors and icons, work practices, ethics and relationships, daily life, learning styles, religion, gender roles, purchasing habits, political influences, customs, and so forth. This is the group that plays a crucial role in defining what the final product looks and feels like when it is finally shipped to the target country.
  - Software engineers and developers—who can (a) define the software architecture based on the target culture's technological advances or limitations, interoperability with other systems, and infrastructure; and (b) develop software that is internationalized and ready to be localized to different target markets. This group must understand how to internationalize a product from the software design perspective.
  - Technical writers and translators—who can (a) write documents with the international audience in mind, so that the text is easily translated into different languages; and (b) translate documents and customize them for the target languages. Translation is a major expense in the localization process. Creating the right documents up front can save time and money in the translation process. Following guidelines on writing text for translation (see Aykin & Milewski, chap. 2, this volume, for more details), identifying clearly what is translated and what is not, and creating a translation process are key elements in translation.

The following section provides high-level guidelines for each of these teams to carry out a smooth internationalization and localization process.

INTERNATIONALIZATION AND LOCALIZATION TEAM ROLES

The three teams (management, marketing, and internationalization/localization implementation) should be able to define strategies and processes together to ensure a smooth flow of information among teams. In this section, I do not cover the roles of management and marketing teams but focus instead on the implementation team, beginning with the key role of the project manager. The project manager is the glue for the entire process and is responsible for getting the localized product out to the target culture on time.

Role of the Project Manager

As the person managing the internationalization and localization process, the project manager has to understand that localization is not just the translation of the documentation and the user interface content. He or she has to help the team understand the extent of internationalization and localization and be a strong advocate for this work. The project manager is responsible for the following:

- A Well-Defined Strategy. Working closely with the management and marketing teams, the project manager must define the product's globalization strategy and localization process. Without a solid strategy, it is not possible to decide whether to go global or not, what level of globalization to aim for, what the target markets are, time to deliver, and what changes are required in the organization structure.
- A Statement of Work. This includes localization methodology, localization vendor information if work is to be outsourced, and a high-level project schedule which includes milestones and budget. The localization vendor can be responsible only for translation, or can include software localization. Adding languages can affect the processes in terms of scaling and re-engineering costs.
- Vendor Management. In most cases, the companies use outside vendors to handle translation. In this case, it is absolutely critical to choose the right vendor, as this work is a key component of the internationalization process. There is a need to have a clear project structure and guidelines for the translators to avoid "garbage in—garbage out." For a project manager, this responsibility includes:
  - Evaluating and qualifying vendors.
  - Establishing corporate pricing structures for translation and localization.
  - Developing long-term partnerships with vendors.
  - Setting clear guidelines on intellectual property and other tools or guides that are generated for the project.
- Communication. Without good communication channels, it is difficult to coordinate activities, reach milestones, and keep budget

and project schedules under control. The communication should be established to ensure a continuous interaction with the entire team. A few suggestions include:

- Scheduled meetings. Nothing beats the face-to-face project meetings where project issues are discussed, goals are communicated, and proactive measures are taken to ensure milestones are met. Depending on the geographic locations of the team members, face-to-face meetings can be replaced with video- or teleconferencing. However, the project manager still needs to go to the other locations to meet with the team members face-to-face. This increases the team's cohesiveness and productivity.

- Online. Build a strong communication infrastructure for teams to share work. A good way to enhance communication and share resources is via company intranets, separate shared links where all documentation and resources can be posted, shared, and discussed.

- Content Management. Manage international corporate content centrally, and leverage the expertise of locals to generate locale-specific content. Communicate with the locale experts regarding new content, changes on the site, branding requirements, user interface guidelines, and so forth. What works for localization of software applications often isn't good enough for web content. Simply translating the words from English is not a good fit for pre-sales and marketing content typically found on web pages. Overall content needs to be reworked for different markets and must "feel right" for a given language. It should be conversational with appropriate style and local expressions. The portions of the content that change when targeted to a specific locale need to be identified. This is achieved by understanding the users of that locale and specifying what works and what the users need. If not done, the customers from the target culture can easily detect content that has been "just translated" rather than created specifically for them.

- A Common Glossary. Developing a central glossary management system prevents using multiple words for the same thing. This facilitates translation, reduces translation cost, and improves the quality of translation. In developing web sites for different target countries, the project manager can choose to make regional teams responsible for maintaining and updating the sites, given that the content and the layout are reviewed by the central office for conformation to company standards including complying with the common glossary.

- Localization Kit. This includes all pertinent information that needs to be communicated to the localization vendor:
  - Product overview
  - Target audience
  - Contact information for the localization managers, engineers, developers, and user interface designers
  - Project description
  - Project scope and metrics
  - Detailed project schedule
  - Specific localization instructions (source and target languages, translation instructions [what and what not to translate], testing instructions [browsers, operating systems], formatting guidelines, etc.)
  - Tools to be used by the localization vendor
  - Directories and files required to build localized versions, and testing tools.
- A Repository of User Interface and Software Elements. It is extremely valuable if there is a central repository that includes company branding tools and guides, all internationalized and localized graphics and icons, reusable code, templates, et cetera.
- An International Quality Assurance Methodology. To ensure delivery of quality products, quality assurance should be carried out throughout the product development and deployment process. This includes defining metrics for quality and methods to carry out quality assurance, including in-country reviews and testing.
- Maintenance of Internationalization and Localization Guidelines. These guidelines are created with input from the internationalization and localization team. They can include international and country-specific user interface design guidelines and software internationalization guidelines.
- A Post-Localization Issues List. This list may include deploying and maintaining updates, analyzing user feedback, and refining processes. For example, managing updates in the content can be a nightmare if a structure is not placed around it. This involves identifying changes, translating only the changes, and incorporating them back into the content.
- Legal Issues. Working with marketing and legal teams to identify any legal limitations, international and national laws, tariffs and customs, tax laws, and export/import structure is crucial to ensuring smooth entry into the target markets. There is a separate section at the end of this chapter that provides high-level guidelines regarding legal issues in internationalization.

Role of the User Interface Designer

A solid background in user interface design is essential here as in any design work, and gathering user requirements before anything else is done must be a top priority. It does not matter what our target culture is, a strong requirements-gathering process can help prevent future problems when going global. We have heard embarrassing anecdotes related to the internationalization efforts of companies. But, we definitely would not want them to happen to us. And more important, we don't want to lose sales because we did not meet the customer needs. Dealing effectively with internationalization and localization process and design issues demands more than just a solid background in user interface design. There are, unfortunately, still user interface designers who do not understand why we are localizing products. These designers may even believe that people from other countries are used to the "American" way and that they like to be "Americanized." This may be true in part: We see kids wearing Hard Rock Café T-shirts in a rural town in Tibet, and we see kids playing Pokeman games (from Japan) in Greece. But what we don't see is kids playing American football in Tibet and people waiting excitingly for the baseball season to start in Greece. It is true that the boundaries are getting grayer every day, but cultural differences do exist and are a key part of usability that designers need to confront. The topic of designing for different cultures is becoming more and more visible at conferences and in publications. Learning from others and reading about cultural differences is a good start. The real understanding comes from going "over there" and observing and collecting data. If that is too expensive, working with in-country people to collect data for you and educate you about their culture regarding design issues will also work.

Here are a few items to consider while designing for different cultures:

- Design with internationalization in mind - If your company has plans to expand its customer base globally, it is important to consider design characteristics that would be acceptable for most countries. It starts from branding your products to developing icons that are acceptable and understandable worldwide. The more design decisions that are truly global, the fewer will have to be localized for individual markets.
  - Icon design: A simple well-known example of a global icon is "mail/e-mail." Instead of having to accommodate different mailboxes/slots around the world, one common icon, "an envelope," solves this design problem. And you do not need to localize it for different cultures. Other examples of commonly accepted and understood icons include books, cameras, movie projectors, clocks, eye glasses, microphones, planes, trains, scissors, no

smoking signs, and so on. Horton (chap. 6, this volume) describes the rules on icon design in detail.

- Religious references: Unless your site is related directly to religion, it is best to avoid religious references. This way, you do not need to worry about who visits your site and who may be offended by it. It is also important to respect the beliefs and practices of different religions. For example, showing uncovered body parts is not appropriate in Islamic countries.
- Culture-specific examples: Referring to culture-specific heroes, symbols, puns, and idioms either may not translate well into other cultures or may be misinterpreted—and both may cause problems. For example, referral to heroes in content is very culture-specific. Presidents Jefferson and Lincoln may be the great heroes in American culture, but cannot replace Mandela in South Africa. Benjamin Franklin is better known as a scientist to the rest of the world, as compared to the United States, where he is better known as one of its first political figures. Similarly, yellow school buses are the transportation vehicles for children going to school in the United States. Other countries have different ways for children to get to school, including walking.
- Cultural preferences: This can include everything from learning styles to color choices. Whereas some cultures may prefer to read the manual before operating a device, others may prefer to talk to sales representatives, and still others may prefer to go with trial and error.
- Data formatting: Choosing the right formatting for the target cultures is the start of any design for international markets. Many times, there may not be any common formatting for all target cultures, and a localized version of the data becomes necessary. In some cases, such as address formats that differ widely from country to country, the designer can still create a generic address format to accommodate most of the world address formats. Aykin and Milewski (chap. 2, this volume) cover many of the data-formatting issues.

- Perform usability evaluations with cultural differences in mind-Early testing of localized versions of the product is very important because testing in one country does not guarantee product usability in other countries. Usability testing uncovers design problems that may need to be corrected in the early stages of development. Susan Dray and David Siegel's chapter on international user studies (chap. 7, this volume) covers a vast amount of information about conducting global usability studies in the trenches of the globe. Anyone who has worked on any global project can tell you stories about how difficult usability studies can be. One thing is clear: You need to accept that any event can go wrong in your studies. The more prepared you are, the less chance you will have for surprises that cause costly delays in the study.

The main principle in global usability testing is to go to the target locale and run the requirements gathering before the design and testing, and perform usability evaluation during and after the design. Of course, this "best case" scenario is more expensive in the short term, but the return on your investment will be tremendous.

Here are some guidelines on global usability testing. (See Dray & Siegel, chap. 7, this volume, for more details.)

- Leverage in-country resources.
- Obtain early and ongoing feedback using a variety of sources of input (customer visits, expert evaluation, user evaluation of prototypes, user trials, etc.).
- Use well-defined processes and modify them when necessary.
- Translate and adapt materials as needed.
- Develop generous schedules; consider local holidays.
- Expect the unexpected (equipment specifications/failure, recruiting, etc.).
- Interpret behavior within the context of the target culture.
- Run the test with local users and in their native language.

Role of the Software Engineer and Developer

Software engineers and developers are the key people who will be implementing all the internationalization concepts and design issues into the product. They are indeed lucky if management calls for a product that must be designed and developed from the beginning with internationalization in mind. Unfortunately, in most cases, the product is an existing one, designed initially with a single country and culture in mind, one with little or no flexibility in the software. Before translating anything, before changing code, and so forth, the software needs to be assessed as to whether it can handle internationalization or not, and whether it can be reengineered to be internationalized. It is important that the software be sufficiently flexible to handle all localization changes for multiple locales. A truly localized product should include:

- One set of source code.
- A single bug-tracking system.

- Isolated localizable resources, including dialogs, macro language, status bars, messages, menus, prompts, toolbars, and sounds.
- Localizable elements, such as time, date, currency, address, names, and so on, that are not hard-coded and should support different character sets.
- Buffers large enough to handle translated text in cases where the text expands.
- Character parsing that is not limited to Latin script.
- Code clearly marked showing what must and must not be translated.
- Content that is presentable in users' language and character set.
- User input that can be received in users' native language and character set.

Role of the Technical Writer and Translator

Technical writers create content that may end up being translated to many languages. In chapter 2 of this volume, Milewski and I cover the general rules on how to write for translation. Besides these general rules, it is very important to pay extra attention to:

- Common glossary/dictionary: For an organization, especially going global, it is very important to create a standard and consistent terminology that is shared by all parties involved in generating content and user interface elements. Placing this standard and consistent terminology in a common glossary is the best way to share among the team members.
- Spell checker: Regardless of the language, a spell checker should be used consistently throughout content generation to ensure correct spelling and use of accents in the target language.
- Culture-specific language: Eliminate all culture-specific metaphors, acronyms, abbreviations, humor, puns, idioms, slang, sports, and other culture-specific analogies from the content to make the content truly internationalized.

A well-written, concise, and culture-free source document serves as a good starting point for translation. If there is a need to include any culture-specific referrals, clear communication among all team members will ensure correct translation and interpretation of these culture-specific examples.

LEGAL ISSUES

Introducing a product into a new country or locale requires more than adapting it to the country's cultural and linguistic requirements. It includes dealing with a variety of legal issues, tax laws, duty fees, value-added taxes, advertisement rules, and warranty policies. Most of these issues are best handled by experts in international and in-country laws and standards. Microsoft (2002) describes their introduction of Windows in mainland China as a good case study for businesses expanding into global markets. Apparently, Microsoft was not aware of the political issues surrounding developing the Simplified Chinese edition of Windows offshore and then importing into the People's Republic of China. Microsoft had to change its strategy after angering the Chinese government and then had to work very closely with the Chinese government and local developers to ensure the success of Windows XP.

Sullivan (2003) reports that the Google search engine has 3.3 billions of text documents indexed as of September 2, 2003. This figure does not include Google discussion posts, image and multimedia files. These documents, discussion posts, and images are from all around the world. NUA (2002) reports that the estimated number of online users in September 2002 was 605.6 million. Not all online users come from a single region or country. Out of this total, there are 190.91 million in Europe, 187.24 million in Asia/Pacific, 182.67 million in Canada and the United States, 33.35 million in Latin America, 6.31 million in Africa, and 5.12 million in the Middle East. Considering these statistics, legal issues become more complicated. Given that there are more than 3.3 billion documents posted from all around the world and 605 million users accessing these documents from all around the world, whose laws will apply?

Ellis (2001) and Aykin (2001) give examples of how cultural differences can have legal implications if one is not careful.

The Internet is full of stories about who is suing whom for damages, copyrights, privacy issues, and so forth. For example, France sued Yahoo for selling Nazi memorabilia on their auction site, and it was not Yahoo's French site that was sued, but its U.S. site. Although the server is located in the United States, the French people could access the site and see the Nazi memorabilia being auctioned there (Aykin, 2001). Hongladarom (2000) lists several guidelines regarding messages on the Internet in Thailand. The guidelines state that messages critical of the King and his royal family are absolutely prohibited, and suggest not posting any messages that contain foul language, sexually explicit content, insults, criticism of any religion, or personal data of any kind.

Finally, the governments of Italy, England, and Scotland have seized computer equipment of anarchist groups that advocate antigovernment violence via the Internet (Schönberger & Foster, 1997).

Currently, there is no single solution, and there is no one legal entity to resolve the legal issues that arise due to the Internet. However, we see some actions now being taken by e-commerce companies, governments, international organizations, and regulatory agencies to bring resolutions to the conflicts raised by the Internet. The following organizations and regulations deal with issues of this type.

The International Organisation for Economic Co-Operation and Development (OECD) (adopted OECD Guidelines for Consumer Protection in the Context of Electronic Commerce (December 1999) for consumer protection in online commerce.)

Global Information Infrastructure Commission *(GIIC)*.

American Bar Association.

Hague Convention on Jurisdiction & Foreign Judgments in Civil and Commercial Matters *(HCCH)*.

E.U. Data Protection Directive.

European Council of Ministers, Brussels 1 regulation.

Safe Harbor Privacy Principles signed by the United States Commerce Department and the E.U. Commission.

These are just a few organizations that are trying to deal with the international issues related to global use of the Internet.

Sinrod (2000) and Ellis (2001) provide a few recommendations on how to protect an entity and its content on the Internet from the laws of various countries around the world:

- *Domain Names:* Under the Anti-Cybersquatting Consumer Protection Act (1999), rightful owners of trademarks or personal names may now bring an action against anyone who, with a bad intent to profit, registers, traffics in, or uses a domain name that is identical or confusingly similar to a mark that was distinctive or famous when the domain name was registered. However, this does not prevent cases of having same domain names in other countries. An example includes a battle over the use of "Bud" or "Budweiser" between St. Louis-based Anheuser-Busch and Czech brewer Budvar. The two companies signed an agreement in 1939 that kept Budvar out of the United States and Anheuser-Busch's Budweiser from using that name in many European countries. In 2000, Budejovicky Budvar introduced their beer to the United States under the Czechvar brand (Realbeer, 2001).
- *Privacy:* Since the explosion of the Internet, privacy laws have gained a lot of attention due to the fact that people enter their names and credit cards online, and that information about people

can be shared by many online providers. People need to trust the confidentiality of the process and the security of their personal data. If your web site collects personal information, there should be a clear privacy policy regarding the handling of the information that is collected. The European Union Data Protection Directive (1995) requires companies that collect information from individuals in Europe to disclose how the information is used. It also requires that individuals be allowed to edit or delete this information (Ellis, 2001). Having a U.S.-based site does not exclude the site owners from this directive. Additionally, the directive bars transfers of personal information from E.U. countries to non-E.U. countries if they do not meet the directive's standards.

- *Arbitration:* A corporate site should include provision for arbitration. There are more than 100 countries that recognize arbitration as a solution to international disputes.
- *Terms of use and legal page:* Including a Terms of Use/Legal page that provides rules about using the site helps avoid and resolve disputes and misunderstandings.
- *Insurance:* Businesses can purchase insurance to cover their liability for claims resulting from online transactions across borders.
- *Tax:* If a company has a physical presence in a country that has a value-added tax (VAT), the company's online transactions, even with customers outside that country's borders, may also be subject to that tax. If the company does not have a presence in the country in which the customer is located, the online transaction may be subject to import taxes. There are a lot of resources to get information on what taxes apply in which countries, including legal entities, shippers, customs brokers, and freight forwarders.
- *Physical location of assets:* If a company has a physical presence in a country, this office is considered to be legally responsible for any issues regarding the company's web site even if the web site server may not be located in that country.
- *Third-party content:* Hosting third-party content or chat lines can become a headache if the content is not monitored carefully. Any offensive material published or advertised becomes the host's liability. For example, eBay was sued by the French government for auctioning Nazi and other racist items to French citizens (Morris, 2000).

There are also government regulations on use of languages in many countries. For example, Quebec in Canada enforces that the language of any published content should be first in French, and other languages can only be introduced after the French version is implemented.

SUMMARY

As was mentioned at the beginning of this chapter, it is very hard to explain all aspects of globalization in one book, let alone in one chapter. This chapter is intended to give a real-world checklist of actions required to successfully globalize a product. In addition, it highlights a few key issues. In the remaining chapters, experts in this field describe the issues in more detail and provide information regarding how differences across cultures and languages affect design.

REFERENCES

Aykin, N. (2001). Legal issues reconsidered: Do the country boundaries matter anymore? In M. J. Smith & G. Salvendy (Eds.), *Systems, social and internationalization design aspects of human–computer interaction* (Vol. 2, pp. 548–552). Mahwah, NJ: Lawrence Erlbaum Associates.

CIO. (2002a). *Metrics: PDA Shipments grow 13.8 percent in Latin America*. October 1, 2002. Source: Gartner Dataquest. Retrieved January 22, 2004, from http://www2.cio.com/metrics/2002/metric445.html?CATEGORY=11& NAME =Globalization

CIO. (2002b). *Metrics: One billion PCs shipped*, July 2, 2002. source: Gartner Dataquest. Retrieved January 22, 2004, from http://www2.cio.com/metrics/2002/metric397.html?CATEGORY=11&NAME=Globalization

CIO. (2003). *Metrics: Internet population reaches 580 million* (source: Nielsen/Netratings), February 27, 2003. Retrieved January 22, 2004, from http://www2.cio.com/metrics/2003/metric508.html?CATEGORY=11&NAME= Globalization

DePalma, D. D. (2002). *Business without borders; A strategic guide to global marketing*. New York: Wiley.

Dunlop, B. (1999). *Success in any language*. March 9, 1999. Retrieved January 22, 2004, from http://www.clickz.com/int_comm/article.php/812791

EU Data Protection Directive. (1995). EU Directive 95/46/EC. *The Data Protection Directive*. 24 October 1995. Retrieved January 22, 2004, from http://www.dataprivacy.ie/6aii.htm

Ellis, R. L. (2001). Web site globalization and the law. *IQPC Seventh International Conference on Web Site Globalization*, San Francisco, March 25–28, 2001.

Hongladarom, S. (2000). *Negotiating the global and the local: How Thai culture co-opts the Internet*. Retrieved January 22, 2004. First Monday, http://www.first monday.dk/issues/issue5_8/hongladarom/

Localisation Industry Standards Association. (2001). *The localization industry primer*. Retrieved January 22, 2004, from http://www.lisa.org/products/primer.html

Microsoft (Dr. International Team). (2002). *Developing international software*. Redmond: WA, Microsoft Press.

Morris, S. (2000). *The importance of international laws for web publishers*. June 2000. Retrieved January 22, 2004, from http://www.gigalaw.com/articles/2000/morris-2000-06.html

O'Hara-Devereaux, M., & Johansen, R. (1994). *Global work: Bridging distance, culture and time*. San Francisco: Jossey-Bass.

Sullivan, D. (2003). *Search engine sizes*. September 2, 2003. Retrieved January 22, 2004, from http://searchenginewatch.com/reports/article.php/2156481

NUA. (2002). *How many online?* Retrieved January 22, 2004, from http://www. nva.ie/surveys/how_many_online/

Schönberger, V. M., & Foster, T. E. (2001). A regulatory web: Free speech and the global information infrastructure. Michigan Telecomm. *Tech Law Review*, 45, University of Michigan Law School. Retrieved January 22, 2004, from http://www.mttlr.org/volthree/foster_art/html

Sinrod, E. J. (2000). *Upside counsel: Know the laws of cyberspace*. Retrieved January 22, 2004, from http://listweb.bilkent.edu.tr/anti-spam/2000/nov/ att-0000/01-upside_counsel.htm

Realbeer. (2001). *A Bud by any other name ....* Retrieved January 22, 2004, from http://www.realbeer.com/news/articles/news-001477.php

Rhinesmith, S. H. (1993). *A manager's guide to globalization: Six keys to success in a changing world*. Alexandria, VA: McGraw Hill.

Chapter 2

Practical Issues and Guidelines for International Information Display

Nuray Aykin
Siemens Corporate Research, Princeton, NJ

Allen E. Milewski
Monmouth University, West Long Branch, NJ

INTERNATIONAL INFORMATION DISPLAY

Internationalization and localization include the study of differences among aspects of locales and how to deal with them. Differences relevant to international design span a wide range. Some differences are abstract and involve conceptualizations of culture and their implications for users' reactions to products and services. Others are more pragmatic and concrete. There are many locales and many dimensions of differences among locales. For the practitioner who is about to tackle an internationalization project, it is daunting to consider the sheer size of the knowledge base required to do a good and insightful job.

This chapter is a guide to a portion of that knowledge base and is intended to be used by the practitioner as a reference during internationalization projects. The chapter focuses on pragmatic guidelines for information display. These are critical, especially in the context of web page design and evaluation. In some design cases, clear guidelines are possible. In other cases, the complexities of international design make straightforward guidelines elusive. In these cases, we resort to providing useful resources and examples that demonstrate the kinds of problems that can be confronted.

We have found it useful, in tackling international projects, to keep in mind the following list of topics and potential problem areas as the top layer of internationalization in any cross-cultural design:

1. *Gr*aphics and icons.
2. *L*anguage: rendering and translating.
3. data *O*bject formatting.
3. *C*olor.
5. *L*ayout.

This, of course, is a grossly incomplete list of concerns for internationalization in general (e.g., content, etc.). But, these items do cover a significant proportion of the typical internationalization problems found in designing web pages (Aykin, 2003). The acronym formed by this list, GLOCL, also provides a useful checklist of problem areas that is easy to remember by recalling that the area of internationalization is motivated by a constant tension between what is *Glo*bal and what is *Loc*al.

GRAPHICS AND ICONS

William Horton (chap. 6, this volume) already covers an extensive array of useful guidelines on graphics and icons that are motivated by cultural differences. These detailed guidelines are not repeated here, but it is worth listing several general rules in using graphics in design:

- Keep the design simple.
- Leave the text out of graphics, or use overlaying text on graphics so that when translated, the icons need not be redone.
- Use universally recognized objects. Examples include using an envelope image for mail instead of a rural mailbox (which is specific to the United States) and using an octagonal stop sign instead of a hand to stop an action.
- Implement different icons for each culture where you cannot find culture-free graphics and icons.
- Be cautious in using images that are potentially sensitive or potentially incorrect. For example:
  - Animals and People—Use of animals as symbols can be offensive to some cultures. For example, the pig is considered a filthy animal in Muslim and Jewish cultures, but it is a symbol of savings in the United States (see coolsavings.com web site).

- Maps—Maps must be drawn based on the latest information. We are in a world of ever-changing borders. It is not acceptable, and is considered as an insult, if a map is not drawn to scale or does not show the current legal borders.
- Directions on how images are read—Not all languages are read from left to right. The same direction in reading applies to displaying images in sequence. In languages where the text is read from right to left, as in Arabic and Hebrew, the images are read from right to left as well.

LANGUAGE: RENDERING AND TRANSLATING

Language is one of the key elements of internationalization. It is probably the element with the highest visibility. It affects everything from how characters are displayed to how layouts need to accommodate potential expansion during translation. One of the complexities here is that languages and countries, while related, do not form a one-to-one relationship. First, many countries can share the same language; for example, English is spoken in some 70 countries. Second, some countries are multilingual, such as Belgium (Dutch, French), Switzerland (Italian, French, German), and Canada (English, French). Finally, we can see different dialects (such as Parisian, Swiss, and Canadian French) and also different writing scripts (see below) in the same language. The lack of a one-to-one relationship between countries, languages, and scripts means that designers must consider both language *and* country as potential determinants of design. The multitude of permutations of language, dialect, country, and script has implications for at least two key aspects of an internationalization project: language rendering and translation.

Language-Rendering Issues

An Example: Japanese and Chinese

Some of the difficulties confronted in rendering languages for user interface displays can be shown with examples of two locales: Japan and China. Japan has one language, but several scripts. All of these Japanese scripts can be used and intermixed in a single sentence.

- *Kanji* is the complete written Japanese language. It has 50,000 Chinese-style characters and can be used for writing a majority of the language (example below from webjapanese, 2001).

一 右 雨 円 火 下 何 外 学 間
気 九 休 金 月 見 五 後 午 語
校 行 高 国 今 左 三 山 四 子
時 七 車 十 出 書 女 小 上 食
人 水 生 西 川 千 先 前 大 男
中 長 天 電 土 東 読 南 二 日
入 年 白 八 半 百 父 分 聞 母
北 木 本 毎 万 名 友 来 六 話

- *Kana*—While Kanji is essentially an ideographic system, Kana are alphabets of written syllables that historically derive from Kanji. Since Kana is simpler than Kanji, they make reading and writing easier. There are two Kana forms:
  - *Hiragana* represents all native Japanese phonemes and sounds. It is used for articles and endings (example below from *Japanese for Otakus*, 2004).

	k	g	s	z	t	d	n	h	b	p	m	y	r	w	n'
a	あ	か	が	さ	ざ	た	だ	な	は	ば	ぱ	ま	や	ら	わ	ん
i	い	き	ぎ	し	じ	ち	ぢ	に	ひ	び	ぴ	み		り	ゐ	
u	う	く	ぐ	す	ず	つ	づ	ぬ	ふ	ぶ	ぷ	む	ゆ	る		
e	え	け	げ	せ	ぜ	て	で	ね	へ	べ	ぺ	め		れ	ゑ	
o	お	こ	ご	そ	ぞ	と	ど	の	ほ	ぼ	ぽ	も	よ	ろ	を	

  - *Katakana* is used to represent foreign names and words other than Chinese or Korean (example below from *Japanese for Otakus*, 2004).

	k	g	s	z	t	d	n	h	b	p	m	y	r	w	n'
a	ア	カ	ガ	サ	ザ	タ	ダ	ナ	ハ	バ	パ	マ	ヤ	ラ	ワ	ン
i	イ	キ	ギ	シ	ジ	チ	ヂ	ニ	ヒ	ビ	ピ	ミ		リ	ヰ	
u	ウ	ク	グ	ス	ズ	ツ	ヅ	ヌ	フ	ブ	プ	ム	ユ	ル		
e	エ	ケ	ゲ	セ	ゼ	テ	デ	ネ	ヘ	ベ	ペ	メ		レ	ヱ	
o	オ	コ	ゴ	ソ	ゾ	ト	ド	ノ	ホ	ボ	ポ	モ	ヨ	ロ	ヲ	

- *Romaji* (Roman Characters) is a set of Roman or English characters. They are used for nontranslated words and numbers. They include:

> Aa Bb Cc Dd Ed Ff Gg Hh Ii Jj Kk Ll Mm Nn Oo Pp Qq Rr
> Ss Tt Uu Vv Ww Xx Yy Zz
> 1 2 3 4 5 6 7 8 9 0

Since characters from all of these sets can be mixed when writing in Japanese, an extremely large number of characters is needed to represent a typical web page in Japan. But although the large number of characters needed to render Japanese web pages is problematic, it is still more manageable than accommodating countries with multiple, regional languages and dialects as well as scripts. Chinese, for example, has two main dialects:

- Mandarin, which is used primarily in Mainland China, and
- Cantonese, which is used primarily in Taiwan.

In addition to the dialects, there are two main scripts:

- Simplified Chinese, which is used in China and Singapore, and
- Traditional Chinese, which is used in Taiwan and Hong Kong.

Character Sets

In order to deal with the processing and rendering of multiple and large scripts, several character sets have been established and standardized. A character set is simply an encoding of some set of language characters into a digital, numeric code. The basic character sets commonly used in software today are as follows:

ASCII. The standard ASCII (American Standard Code for Information Interchange) character set consists of 128 characters ranging from 0 through 127 with the first 32 consisting of nonprinting characters. The rest is assigned to letters, numbers, punctuation marks, and some special characters. The extended ASCII character set includes the standard ASCII plus characters from 128 through 255 representing additional special, mathematical, graphic, and foreign characters.

The ASCII character set supports only English, Swahili, and Hawaiian languages.

The ISO-8859 Family of Character Sets. Latin-1 (ISO 8859-1) is the default character set for the Web. However, it is sufficient only for the European languages. It uses 8-bit encoding and permits a maximum of 256 characters. The ISO 8859 Character Set includes language-specific groupings:

ISO 8859-1 [Latin 1] Western Europe (French, Italian, German, Spanish, ...)

ISO 8859-2 [Latin 2] Central/Eastern Europe (Hungarian, Polish, Romanian, Croatian, ...)

ISO 8859-3 [Latin 3] Northern Europe (Esperanto and Maltese)

ISO 8859-4 [Latin 4] Southern Europe (Estonian, Latvian, Lithuanian, Greenlandic, and Lappish)

ISO 8859-5 [Cyrillic]

ISO 8859-6 [Arabic]

ISO 8859-7 [Greek]

ISO 8859-8 [Hebrew]

ISO 8859-9 [Latin 5] Turkish

ISO 8859-10 [Latin 6] Nordic (Removed Latvian from 8859-4 and added Icelandic)

ISO 8859-11 [Thai]

ISO 8859-12 [unassigned]

ISO 8859-13 [Latin 7] Baltic Rim (Estonian, Latvian, Lithuanian)

ISO 8859-14 [Latin 8] Celtic (Gaelic and Welsh)

ISO 8859-15 [Latin 9] Latin 0 (Removed little used characters from 8859-1 and added Euro €)

Unicode (ISO 10646 BMP).   Unicode is a large character set that includes most of the world's languages. It uses 16-bit encoding and allows 65,000 characters. This form is called UCS-2 (Universal Character Set 2-bytes). Unicode is the first level of ISO 10646, and it is also called a BMP (Basic Multilingual Plane) or Plane Zero. More information on Unicode can be found at www.unicode.org.

UTF-8 (Universal Character Set Transformation Format).   UTF-8 is an addendum to ISO 10646 and provides compatibility with ASCII. The ASCII characters are represented by 1 byte (8 bits) rather than 4 bytes (32 bits). UTF-8 format has the unique ability to preserve the full ASCII range.

Consequently, UTF-8 is compatible with file systems, parsers, and other software that rely on US-ASCII values.

Fonts

Character sets are ordered collections of abstract references to characters and do not specify their exact visual appearance. On the other hand, fonts are collections of glyphs of a specific style and appearance. A single character set can be associated with, or used by, one or more fonts. Text must be written in a font that supports the corresponding character code set being used. Characters, especially non-Latin characters may require different line heights, spacing between lines, spacing between characters, character widths, and screen resolution. Some characters, such as those in Japanese, may not require ascenders, descenders, or capitalization. Below are guidelines related to fonts when dealing with different characters sets:

- Provide enough space between lines to accommodate changes in line heights, or automatically adjust line heights depending on the characters used.
- Provide enough interline spacing to ensure clear separation between lines and to accommodate underlining.
- Do not use overly ornate or decorative fonts because accents and special characters can become illegible.
- Choose fonts that can accommodate target languages. For example, Helvetica and Times can provide accents. But Helvetica CE or Times CE should be used to support the characters in central European languages.
- Do not assume the availability of standard fonts.
- Be prepared to support mixed (multinational) formats.
- Test your fonts for different target languages with different browsers.
- Most non-Latin languages require proportional spacing. For example, Arabic cannot be implemented on a monospaced display.

Text Direction

Not all languages read from left to right. There are three types of text directionality: left-to-right, bidirectional, and vertical. The most common is the left-to-right direction that we see in Latin, Cyrillic, Greek, Thai, and Indic languages. Chinese, Japanese, and Korean use left-to-right and vertical text. Hebrew and Arabic are bidirectional (Bi-Di) where Hebrew and Arabic characters are displayed right to left, and numbers and Latin characters are displayed left to right. Arabic and Hebrew are typically read from right to left; however, words using non-Arabic characters and Arabic numbers are read from left to right. The following example is from Microsoft's Saudi Arabia Web site http://www.microsoft.com/middleeast/saudi/campaigns/msdn.asp. The text shows how the bidirectional text is displayed. Note that Arabic characters are right justified, whereas left-to-right directional languages, in non-Arabic text, are left justified.

فقـط الوقـت والـدعم والثقـة اللازمـة لتجـاوز حـدود المطـور العـادي، بـل MSDN لا يوفـر الاشــتراك لــدى

قبـل أي شــخص آخــر، والــتي لا تقــدر بثمـن عنـد Visual Studio®.NET يوفـر لـك إصـدارات بـرامج مثـل

VS.NET (2003) تطويـر التطبيقــات الحديثــة والرائــدةكـن الأول الـذي يحصـل علـى الإصـدار القـادم مـن

قائيــافـي بضـعة شــهور تـل

Order may also be dependent on context. For example, consider the string "123-4567" in an Arabic context. If it is a telephone number, the reader reads it as: 123-4567. If the string is a subtraction, the reader reads it as subtracting 123 from 4567.

When designing a page or screen layout, it is important to consider whether different text direction may need to be accommodated in the final product. The design should be flexible enough to support this functionality of text direction.

Paper Size

Printing onto paper is another aspect of language rendering. Except in North America, most countries use the paper sizes specified by ISO 216, which is based on the metric system. Therefore, it is important to allow the right amount of margin if documents are shared across the globe; otherwise, printing could be a problem because printers do not always supply both U.S. and ISO 216 types of paper. Although Japan has adopted the ISO 216 paper sizes, it still has different standards for the sizes of folders, defined in the JIS P 0138-61 standard (Kuhn, 2002). Table 2.1 shows the most commonly used paper and envelope sizes defined by the United States and ISO 216.

TABLE 2.1
Paper Sizes in the United States and ISO 216

United States		ISO 216	In metric units	In inches
A (letter)	8½ × 11 in.	A4	210 × 297 mm	8.27 × 11.69 in.
Legal	8½ × 14 in.			
B (ledger)	11 × 17 in.	A3	297 × 420 mm	11.69 × 16.54 in.
Business Card	2 × 3½	A8	53 × 74 mm	2.07 × 2.91 in.

Hardware Concerns

A final set of language rendering concerns in internationalization has to do with display hardware. There are several key points to consider when you are dealing with text entry and display hardware from different regions or countries.

- Systems must be able to understand and display the character set used by the keyboard. Make sure all required characters can be entered by testing your system with different locale keyboards.
- Different keyboard layouts may be used in different countries. For example, many European countries use an AZERTY keyboard instead of a QWERTY keyboard.
- Each country has specific power, signaling, labeling, and safety requirements.
- Make sure the printer has the fonts to display your coded-character set.
- Make sure the printer can handle different paper sizes (e.g., A4). Or your page cannot be printed without losing text in the margins due to different paper sizes.

Language Translation Issues

Language translation is an essential component of globalization and is one of the most expensive. One way to keep translation costs down is to provide translation tools. For example, Microsoft (2002) defines its localization kit as a subset of tools, source files, and binary files that can be used to create a localized edition of a program. The localization kit is generally given to translators or third-party contractors. Having a localization kit helps maximize the accuracy of translation and minimize the support the translators need from the team.

In regard to translation, it is important to:

1. Identify the content of strings in the code to be translated. This is done by either including comments in resource files or providing separate text files to the translators.
2. Provide clearly written text because translators will often have difficulty translating ambiguous terminology.
3. Provide translation glossaries. A common glossary will improve consistency across translators and across product lines.

A strong team of technical writers is important in generating content ready for translation. The quality of translation is highly correlated with how well and clearly the original text is written. Having the content prepared correctly from the beginning saves costs and time during the translation process, but it requires careful preparation. Even just making sure that there are no unnecessary words used in writing would save money, given that most translators charge per word. In her book, Hoft (1995) provides detailed guidelines for technical writers who develop content for an international audience.

Abbreviations and Acronyms

Abbreviations and acronyms can be a headache for translators. They should be included in the translator's glossary so that the meaning does not get lost during translation. They can be spelled out, translated, or left unchanged if they are universal (e.g., ISO, WHO). The full explanations for acronyms and abbreviations should be available to the user in a glossary.

Spelling

The same language used in different countries can have different spelling rules. For example, *neighbor* in the United States is spelled as *neighbour* in U.K. English. Similarly:

English (U.S.)	English (U.K.)
internationalization	internationalisation
program	programme
color	colour
center	centre

It is important that the users feel the language is theirs and not an imported version from another country. This means that content prepared for one locale should be reviewed for appropriateness before use in another locale that speaks the same language.

Text Expansion

When translating from English to other languages, text can expand as much as 30% to 200%. Hoft (1995) states that most translation companies recommend that the layout should be designed to accommodate a 30% text expansion rate. This 30% rate applies to text that is presented in paragraphs (over 70 characters). If the text is less than 10 characters long (e.g., labels, field names, or icons), the expansion rate can go as high as 200%, sometimes reaching 400%. Another common rule is that a double-byte character occupies twice as much physical space as a single-byte character. Some examples of text expansion are shown below:

News (English) <-> Haberler (Turkish)

Product Support (English) <-> Produktunterstützung (German)

Search (English) <-> Búsqueda (Spanish)

Undo (English) <-> Ongedaan (Dutch)

In order to accommodate possible text expansion due to translation, the following guidelines are useful:

- Provide flexibility in the screen/page layout to accommodate different size text and fonts.
- Leave room to accommodate text expansion. Hoft (1995) suggests a few ways to do this:
  - Use a larger type size for the source language and a smaller size for the translated version. (Be certain, however, that readability is not sacrificed.)
  - Increase the width of margins on the source document when text expansion occurs after translation.
  - Consider kerning, leading, and changing the space between paragraphs. This reduces the space between lines.
- Place labels above the fields when the screen real estate permits. Placing labels above the entry fields provides flexibility in case the length of the label expands after translation. Using wraparound labels is another solution if the screen real estate is a problem.
- Use symbols and icons rather than text. When appropriate, using graphical representation can reduce the cost of translation and reduce the number of changes to page layout.
- Use tabs to reduce the amount of information on a page.
- Use automatically expanding text objects.
- Consider alternative layout styles.
- Allocate enough memory to allow for text expansion.

Sorting

Languages also differ in the ways characters are sorted. The sorting rules for different languages should be able to handle:

- Accents.

- Combinations of characters (e.g., in Spanish, "cho" comes after "co" because "ch" is treated as a separate character that comes after "c" in the alphabet; also, in German the character 'ß' is sorted as "ss").
- Upper- and lowercase differences (in many languages uppercase is sorted after lowercase).
- Non-Latin Scripts (e.g., Hebrew and Arabic).
- Far-Eastern languages (Japanese, Chinese, Korean).

The ISO/IEC 14651:2000 standard titled "International String Ordering" provides universal default collation orders for multiple languages (available online at http://www.iso.org/iso/en/cataloguedetailpage.catalog uedetail?csnumber=250).

Writing Practices

The following guidelines, summarized from the GNOME Documentation Style Guide (2003) and Hoft (1995), concern writing practices that can make the translation easier and more cost effective.

Sentence Structure.

- Use short sentences and substantial white space.
  Instead of: If you would like to turn the computer off, you can select the Shut Down item from the Start menu.
  Use: To turn the computer off, choose Shut Down from the Start Menu.
- Do not use joined sentences (*and/or/so*)—use one thought per sentence.
  Instead of: This menu item allows you to set the page properties, so you can define your own page formatting.
  Use: This menu item allows you to set the page properties. You can define your own page formatting.
- Use active prose.
  Instead of: User privileges are set by the system administrator.
  Use: System administrator sets the user privileges.
- Choose formal vocabulary.
  Instead of: I will pack my bag.
  Use: I will prepare my luggage.
- Avoid gerunds (*-ing* form).
  Instead of: Clicking File opens the File menu.

Use: Click File to open the File menu.

- Avoid indefinite pronouns (e.g., *this, it, that, those*).

Instead of: The Spell Checker identifies misspelled words. It is activated by selecting "Spell Checker" from the Tools menu.

Use: The Spell Checker identifies misspelled words. To activate, select "Spell Checker" from the Tools menu.

- Use present tense.

Instead of: Clicking on Help will activate the Help menu.

Use: Click Help to activate the Help menu.

- Avoid *there is/are*.

Instead of: In the menu, there are five selections.

Use: The menu contains five selections.

- Use parentheses to introduce abbreviations, not to provide explanations.

Instead of: Place the localizable elements in separate files (known as resource files).

Use: Place the localizable elements in separate files, known as resource files.

- Using pre-organizers such as graphics, icons, and lists helps the user as well as the translator.

Instead of: File menu contains New, Open, …

Use: File menu contains:

- *New*
- *Open*
-
-

- Avoid semicolons.

Instead of: The first item in the menu is available; all the others are grayed-out.

Use: The first item in the menu is available. All other menu items are grayed-out.

- Avoid personal pronouns (other than *you*).

Instead of: We suggest using drag-and-drop to move items.

Use: Use drag-and-drop to move items.

- Use *can* or *might* instead of *may*. Use *may* only for granting permission.

Instead of: These options may be selected.

Use: You can select these options.

- Replace contractions with complete words.

Instead of: If you're available.

Use: If you are available.

Terminology.

- Eliminate culture-specific metaphors.
- Avoid acronyms and abbreviations.
- Avoid jokes, humor, and idioms (all are culture-specific and may be offending or difficult to understand and/or translate).
- Avoid gender-specific references.
- Avoid colloquial language.
- Use standard and consistent terminology throughout a document. Create a glossary and permit one definition per item. Using one word for the same object/event eliminates translation problems and provides one-to-one matching for that word. The example in the next list item shows the dangers of using different terminologies for the same thing.
- Consider terminology differences between countries using the same language. For example, wall *outlet, socket,* and *electrical recepta-cle* in the United States can mean 'plug socket,' 'electrical point,' and 'power point' in the United Kingdom. On the other hand, in the United Kingdom, *receptacle* can mean a 'container' or 'trash bin,' whereas *outlet* can mean 'exit.'
- Use a spell checker to ensure correct translation and spelling.
- Ensure that the translated version retains the technical accuracy of the original language version.
- Test your translators by retranslating the sections of the document back to its original language.

DATA OBJECT FORMATTING

Formatting of data objects and fields is another problem area for internationalization. Here are some issues and guidelines for the data objects most often encountered in web site design.

Date and Time

Date formats are sometimes associated with specific languages or locales. Although ISO 8601 specifies a standard for date and time presentation, many countries keep their own date and time formatting instead of adopting the international standards. However, this international standard is a good way of keeping internal representation of date within a software application, and helps to synchronize servers around the world.

The international standard date notation is:

YYYY-MM-DD

where YYYY is the four-digit year, MM is the two-digit month of the year between 01 (January) and 12 (December), and DD is the two-digit day of the month between 01 and 31. For example, 1988-01-11 represents the 11th of January in 1988. This date can be seen in various notations in around the world: 1/11/88, 11/1/88, 88/11/1, 11 January 1988, January 11 1988, and so on. The delimiter to separate the year, month, and day can be a forward slash (/), space, hyphen (-), or period (.).

Many software platforms, tools, and languages such as Java and .NET provide how-to's and resources for presentation of dates and times in different languages.

The differences in date formatting can be observed in the long or short date representations. Dates are most commonly displayed as day, month, and year. In some countries, the month may be displayed before the day and the year; in other countries, the year may be displayed before the month and the day.

Here are the proper formats for some key locales:

MM DD YYYY - United States

DD MM YYYY - Most European countries, Australia, Canada

YYYY MM DD - South Africa, Japan, Korea, China, Taiwan

Table 2.2 shows long and short date formats for different countries.

In long date formatting, the month is usually spelled out. Using a three-letter abbreviation for month may work well for English-speaking countries, but can cause problems in other languages. For example, the months June and July are *Juin* and *Juillet* in French. If the three-letter abbreviation is forced, then both months could be abbreviated as *Jui*.

The following are guidelines on how to treat date formatting for different locales:

- Use the correct date representation for the chosen locale.
- Make use of the properties of the tools, platforms, and languages in providing support for date and time presentations.
- If you must choose only one date format for display for use around the world, use dd Month yyyy (e.g., 20 August 2000). Even if it is not formatted correctly for a given locale, the content is unambiguous.
- Make provisions to accommodate different delimiters for date, month, and year. Include a forward slash (/), period (.), space, or hyphen (-) as options.

TABLE 2.2
Long and Short Date and Time Formats
for Monday April 14, 2003 in Different Countries

Language/Country	Long Date Format	Short Date Format	Time Format
Arabic – Saudi Arabia	1424 ,صــفر12	1424/02/12	ص 02:34:17
Chinese – People's Republic of China	2003年4月14日	2003-4-14	14:34:17
Chinese-Singapore	星期一, 14 四月, 2003	14/4/2003	PM 2:34:17
Dutch-Belgium	maandag 14 april 2003	14/04/2003	14:34:17
Dutch-Netherlands	maandag 14 april 2003	14-4-2003	14:34:17
English– United Kingdom	Monday 14 April 2003	14/04/2003	14:34:17
English – United States	Monday, April 14, 2003	4/14/2003	2:34:17 PM
Finnish-Finland	14. huhtikuuta 2003	14.4.2003	14:34:17
French– France	lundi 14 avril 2003	14/04/2003	14:34:17
French-Switzerland	lundi, 14. avril 2003	14.04.2003	14:34:17
German – Germany	Montag, 14. April 2003	14.04.2003	14:34:17
Greek-Greece	Δευτέρα, 14 Απριλίου 2003	14/4/2003	2:34:17 μμ
Hebrew – Israel	שני 14 אפריל יום 2003	14/04/2003	14:34:17

TABLE 2.2 (continued)

Hindi – India	14 अप्रैल 2003	14-04-2003	14:34:17
Italian – Italy	lunedì 14 aprile 2003	14/04/2003	14:34:17
Japanese – Japan	2003年4月14日	2003/04/14	14:34:17
Russian – Russia	14 апреля 2003 г.	14.04.2003	14:34:17
Spanish–Mexico	Lunes, 14 de Abril de 2003	14/04/2003	02:34:17 p.m.
Spanish–Spain	lunes, 14 de abril de 2003	14/04/2003	14:34:17
Swedish–Sweden	den 14 april 2003	2003-04-14	14:34:17
Turkish–Turkey	14 Nisan 2003 Pazartesi	14.04.2003	14:34:17

- Do not assume a three-character space for month and day abbreviations because this may not work in some languages.

Time

As ISO 8601 specifies, the international standard notation for the time of day is

hh:mm:ss

where hh represents the hours (00–24), mm represents the minutes (00–59), and ss represents the seconds (00–60).

Many countries use the 24-hour time notation (e.g., 15:30) rather than 12-hour AM and 12-hour PM notation (e.g., 3:30 PM). In the United States, the 24-hour notation is referred to as "military time." The delimiters to separate hours, minutes, and seconds can be a colon (:) or a period (.). Table 2.2 also includes the time formatting for different countries.

ISO 8601 has been adopted as European Standard EN 28601 and is therefore now a valid standard in all E.U. countries, and all conflicting national standards have been changed accordingly (Kuhn, 2001).

Calendar/Holidays/Start of Week

There are a few aspects regarding calendars that we need to consider when designing for the global markets.

- Calendar Issues: Although the Gregorian calendar is used in almost every country, there are still cases in which a different calendar might be considered. These include:
  - Arabic calendar (Note: In Islamic countries, the date changes at sunset, not midnight)
  - Jewish calendar
  - Iranian calendar
  - Japanese Imperial calendar
- Holidays: It is important to know the holidays of the target country, especially if you are doing business with them. Do not assume that everyone celebrates Christmas. Many non-Christian countries do not have a Christmas holiday. Even new year celebrations may vary. The Chinese New Year does not coincide with the new year associated with the Gregorian calendar.
- Start of the week (Monday vs. Sunday): In the United States, the week begins on Sunday. In most other parts of the world it begins on Monday. For applications that deal with scheduling, this can be an issue. The user interface showing weeks for scheduling may look different for the U.S. users than for the European users.

Numeric Formatting

Number formatting differs from country to country. The elements of numeric formatting includes:

- Thousands and decimal separator: Separators can be a character or a space. Table 2.3 shows different separators for numeric representation.
- Number of digits between separators: The number of digits between separators can vary from country to country. For example, 123,456,789.00 in the U.S. can be shown as 12,34,56,789.00 in Hindi.
- Negative numbers: Negative numbers may be indicated by using a minus sign before or after the number, or by enclosing the number in parentheses or brackets.

-123

123-

(123)

[123]

TABLE 2.3

Numeric Formatting and Currency Representation for Different Countries

Language/Country	*Numeric Formatting*	*Currency*
Arabic - Saudi Arabia	123,456,789.00	ر.س. 123,456,789.00
Chinese - PRC	123,456,789.00	¥123,456,789.00
Chinese-Singapore	123,456,789.00	$123,456,789.00
Chinese-Taiwan	123,456,789.00	NT$123,456,789.00
Dutch-Belgium	123.456.789,00	123.456.789,00 €
Dutch-Netherlands	123.456.789,00	€ 123.456.789,00
English- United Kingdom	123,456,789.00	£123,456,789.00
English - United States	123,456,789.00	$123,456,789.00
Finnish-Finland	123 456 789,00	123 456 789,00 €
French- France	123 456 789,00	123 456 789,00 €
French-Switzerland	123'456'789.00	SFr. 123'456'789.00
German - Germany	123.456.789,00	123.456.789,00 €
Greek-Greece	123.456.789,00	123.456.789,00 €
Hebrew - Israel	123,456,789.00	₪ 123,456,789.00
Hindi - India	12,34,56,789.00	रु 12,34,56,789.00
Italian - Italy	123.456.789,00	€ 123.456.789,00
Japanese - Japan	123,456,789.00	¥123,456,789
Korean-Korea	123,456,789.00	₩123,456,789
Russian-Russia	123 456 789,00	123 456 789,00p.

(continued on next page)

TABLE 2.3 (continued)

Spanish–Mexico	123,456,789.00	$123,456,789.00
Spanish–Spain	123.456.789,00	123.456.789,00 €
Swedish–Sweden	123 456 789,00	123.456.789,00 kr
Turkish–Turkey	123.456.789,00	123.456.789,00 TL

Names and Addresses

In the world of e-business, there are many formats for address forms. In a typical correspondence, an address consists of

- Title
- First/given name
- Last Name/Family Name/Surname
- Street number, building number, street name
- City or town
- State/province/region
- Country
- Zip or postal code (can include alphanumeric characters)

Labels given for each of the these address entries may vary from country to country. Here are sample address formats for Mexico and the United States.

Mexico

Name (Paternal Name, Maternal Name, First Name)

Street and Number

Building, Floor, Suite #

Colony

City, State

Postal Code

United States

Name (First Name, Middle Initial, Last Name)

Number and Street

City, State Zip Code[1]

Differences in address formatting could be seen in the labels of the name fields. For example, *First Name* is mainly used in the United States, whereas most of the world uses *Given Name*. *Last Name* in the United States becomes a *Family Name* or *Surname* in other countries. And having multiple names or last names adds complexity to the problem. For example, people in Mexico have many names, including two last names.

Titles are commonly used in addresses outside the United States. Even in personal letters, the use of a title shows a level of respect for the addressee. In addition to Mr./Mrs./Ms., the titles Dr. and Professor are extremely important in some cultures. For example, the corporate directory for AT&T, a U.S.-based company, includes names and job titles. The corporate directory for Siemens, a Germany-based company, includes names, job titles, and professional titles.

The other fields in the address format include the street, building, town, city, country, zip/postal code, and so on. The formats for these fields vary from country to country. The differences could be in the number of address lines, order of address lines, number of names (first and middle initial and last/given/surnames), street addresses, level of hierarchy in terms of town, city, state/province/region, and country, and of zip/postal codes. For each of these fields, there should be no assumption on format and length. For example, in the United States the zip code consists of only numeric values; however, in many parts of the world it might include alphanumeric entries and might be longer than a 5-digit zip code.

However, there are many attempts to generate a generic address form so that there is no need to create a different address form for each country. This is especially important for online registration/shopping/ordering practices. Many times, it is impossible to identify which address format to use for each request. Therefore, it becomes very meaningful to provide a generic address format which spans multiple locales.

Figure 2.1 shows a good example (from www.amazon.com) of a generic address format that can handle many of the different address formats in the world.

---

[1] For more examples of country addresses, see Microsoft (2002), which provides different address formats for a set of selected countries.

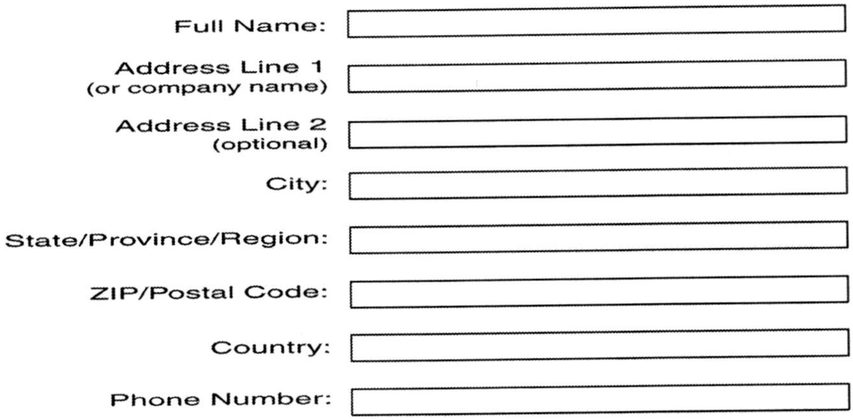

Full Name:	
Address Line 1 (or company name):	
Address Line 2 (optional):	
City:	
State/Province/Region:	
ZIP/Postal Code:	
Country:	
Phone Number:	

FIG. 2.1.   Sample universal address format.

Telephone Numbers

Like other formatted fields, the format for telephone numbers varies from one country to another. The differences between telephone numbers are:

- The total number of digits (varies from six to fifteen)
- Separators (e.g., hyphens (-), periods (.), parentheses (( )), and spaces)
- Groupings of numbers (two to six digits per group)
- Country codes (one to three digits)
- Long distance access code to access national or international calls (could be 00, 011, 001, 0, etc.)
- Code to access an outside line on a business phone (in the United States, "9" is commonly used to place a call outside the office)
- Extensions (some phone numbers include extensions that should be taken into consideration during design).

Table 2.4 provides a few examples of different telephone number formatting:

In addition to different formats in telephone numbering, in some countries, letters are not used on the keypad. This can make it difficult to dial numbers that are given as mnemonics, as in 1-800-WEATHER. Radio Shack used to sell templates for travelers to use on telephones without letters.

ITU-T *Recommendation E.164* (The International Public Telecommunication Numbering Plan; 1997), defined by the International Telecommunications Union (ITU), formerly Comité Consultatif International de Tele-

TABLE 2.4

Sample Telephone Number Formats

Australia	649-800-445-768
Austria	1234 56 78 90
Belgium	12-345 67 89
Denmark	12 34 56 78
Germany	(123) 4 56 78 90
Italy	(12) 3456789
Portugal	123-456 78 90
Japan	123-45-6-789-0000
United States	(123) 456-7890

graphique et Telephonique (CCITT), details the format and structure of the Public Switched Telephone Network (PSTN) numbering plan. Within this Recommendation, it was proposed that all switches and networks should have the capability of handling international telephone numbers with a maximum length of 15 digits. However, given that the separators (e.g., spaces or parentheses) can end up being part of the telephone number entry, the length of the telephone numbers may exceed 15 digits. It is important not to limit the digits in a phone number entry in fields and also not to parse the telephone numbers into area codes and phone numbers as there may not be an area code, but only a city code in some countries. Users should also be able to enter telephone numbers in free format (can use (,), -, /). Consider these notations in determining field sizes.

Currency

There are many different currencies in the world. Table 2.3 shows a few examples of currency representation. Although some currencies have special symbols (such as U.S. dollar as $, Japanese Yen as ¥, British Pound as £), many use abbreviations of the currency name (such as TL for Turkish Lira). ISO 4217 (Codes for the Representation of Currencies and Funds) defines three-letter abbreviations for world currencies.

The general principle used to construct these abbreviations is to take the two-letter country abbreviations defined in ISO 3166 (Codes for the Representation of Names of Countries) and append the first letter of the currency name (e.g., GBP for the Great Britain Pound, and USD for the United States Dollar). In the case of currencies defined by supra-national entities,

ISO_4217 assigns two-letter entity codes starting with "X" to use in place of country codes (e.g., XCD for the Central Caribbean Dollar; Allen & Hall, 2003). Similarly, because there is no specific country associated with Euro, the code for Euro is EUR, with a well-recognized symbol, €.

It is common practice to use the symbols for well-known currencies (such as $ or £) when monetary amounts are displayed for an international audience. For local audiences, it is again a practice to use the currency symbol since the users are familiar with their own currency symbol. However, it is advisable to use the three-letter code if the audience is international and the symbol is not well recognized worldwide.

Monetary Values

The format for representing monetary value differs from country to country. Most locales use their numeric formatting rules in terms of a decimal and thousands separator. But this is not always the case. For example, in Estonia, the period is used as the decimal separator for Estonian Kroons (123.45 kr), but the comma is used as the decimal separator for other numeric representations (123,45; Microsoft, 2002).

Placement of Currency Symbol and Negative Sign

The placement of the currency symbol and the negative sign also differs from country to country. Some countries, such as United States, use parentheses to indicate negative monetary value.

U.K.	–£ 1,324.55
U.S.	($1,324.55)
Netherlands	1,324.55 € –
France	–1,324.55 €

Sizes and Measurements

The United States uses the imperial system for the measurements, whereas most countries use the metric system. For computer applications, this is especially important for files that rely on measurements (e.g., page size, icons, clip art) that are shared across regions. Sizes and measurements that may change from country to country include:

- Typographic units (point, pixel, measurement units).
- Temperature measurements (Fahrenheit or Celsius)

- Paper and envelope sizes (see page 28)
- Clothing sizes

Metric Versus Imperial Units

The Imperial system uses the foot, the pound, and the Fahrenheit (F^o) scale to measure length, mass (weight), and temperature, respectively. The metric system uses the meter, the kilogram, and the Celsius (C^o) scale.

Most of the world uses the metric system, except for Burundi, Liberia, Yemen, Rwandese Republic, Union of Myanmar, and the United States of America.

Typographic Units

Vekulenko (2000) and Kuhn (1999) summarize the differences between the typography units used around the world. To name a few widely used ones:

- 1 point (Didot) = 0.3759 mm = 1/72 of a French Royal inch (27.07 mm) = about 1/68 inch
- 1 point (ATA) = 0.3514598 mm = 0.0138366 inch
- 1 point (TeX) = 0.3514598035 mm
- 1 point (Postscript) = 0.3527777778 mm = 1/72 inch
- 1 point (l'Imprimerie nationale, IN) = 0.4 mm

When sharing images between different locales, be certain to use the same point system to avoid design mistakes.

Temperature

Before the 1970s, most English-speaking countries used the Fahrenheit scale. Since then, most countries adopted the Celsius temperature measurement scale, except in the United States, which uses the Fahrenheit scale. Many travel sites now show temperatures in both scales to help travelers.

Clothing Size

It is especially important for online merchandise catalogs to show metric and imperial shoe and clothing sizes. Even if the target market is the United States, there are many people in the United States from different countries who will be pleased to see the sizes given in the system that they are used to. Many travel agencies' sites, such as American Express, provide clothing size charts to show the conversions for different countries. A useful bit of information for travelers! Table 2.5 shows the conversion for clothing sizes.

COLOR

Color has very different meanings around the world and therefore it is a significant element, not just for computer user interfaces, but for all forms of international interaction. Table 2.6 shows different meanings of colors for various cultures (adapted from HP, 2003; Morton, 2003; Prime & Wilson, 2002; Jinsheng, 2001). For example, Holzschlag (1999) and Morton (2003) point out that the color purple represents death in Catholic parts of Europe and that Euro Disney should have done their homework well before choosing purple as a signature color.

TABLE 2.5
Clothing Sizes for the United States, Europe, and the United Kingdom

Women's Dress Sizes			Women's Shoe Sizes		
United States	Europe	United Kingdom	United States	Europe	United Kingdom
6	36	8	5	35	3.5
8	38	10	5.5	36	4.0
10	40	12	6.0	36	4.5
12	42	14	6.5	37	5.0
14	44	16	7	37	5.5
16	46	18	7.5	38	6.0
18	48	20	8.0	38	6.5
20	50	22	8.5	39	7.0
22	52	24	9.0	39	7.5
Men's Suit Sizes			Men's Shoe Sizes		
United States	Europe	United Kingdom	United States	Europe	United Kingdom
36	46	36	8	41	7
38	48	38	8.5	42	7.5
40	50	40	9.5	43	8.5
42	52	42	10.5	44	9.5
44	54	44	11.5	45	10.5
46	56	46	12.0	46	11

TABLE 2.6
Color Meanings in Various Cultures

Color	Meaning for Different Cultures
Red	China: Good luck, prosperity, happiness, marriage Japan: Anger, danger Middle East: Danger, Evil India: Purity, life Egypt: Death South Africa:Mourning USA and Western Europe: Danger, anger, stop
Blue	China: Immortality, heavens, depth, cleanliness Japan: Villainy India: Color of Krishna USA and Western Europe: Masculinity, calm, authority, peace
Yellow	East Asia: Sacred, Imperial, Royalty, Honor Middle East: Happiness, prosperity India: Religious color (celebration),success Egypt: Mourning USA and Western Europe: caution, cowardice
Green	Asia: Family, harmony, health, peace, life, youth, energy, growth Muslim countries: Religious color, fertility, strength Ireland: National color USA and Western Europe: Safe, go
White	Asia: Mourning, death, purity Middle East: Mourning, purity India: Death, purity USA and Western Europe: Purity, virtue
Black	Asia: Evil influences, knowledge, mourning Middle East: Mystery, evil USA and Western Europe: Mourning, elegance (black dress and tuxedo), death, evil
Brown	India: Mourning
Purple	Thailand: Mourning Catholics: Death Asia: Wealth, glamour Western Europe: Royalty
Orange	Ireland: Religious significance China: Light and warmth USA: Inexpensive goods

There are numerous excellent resources that deal with color usage. Jill Morton's colormatters.com web site has extensive information on color studies and online books on color. Spartan (Multimedia Productions,1999) gives many examples related to different cultures in their association with colors. Here are a few examples from Spartan's colorful multimedia presentation:

- In the United States, Americans tend to relate blue to authority, whereas in Canada, red is authority, as well as an inspirer of nationalism. In the British Isles, black is commonly used for authority figures. The use of red and black in Britain is usually reserved for honorary positions, such as the Royal Guard.
- White is a very common color in hot Mediterranean countries, such as Greece. The use of white dates back to ancient Greece when all buildings, sculptures, and so on were required to be clean white.
- In China, red and gold are used for joyful, festive occasions. They are also commonly used for religious shrines and temples, as well as other forms of architecture.
- Orange is widely used by Indians in their attire and as a means of attention. It also has religious significance in Hindi and Buddhism.
- The Japanese spectral range of colors is mainly muted, pastel types, with the exception of bright red.
- In Korea, light blue and other pastel colors in clothing are usually reserved for the wealthy and for festivals. Commoners usually wear white or grays.
- In many African cultures, black is a symbol of strength, and white is a symbol of fertility.
- Native Indians of Mexico used bright colors for their ceremonial costumes. Possibly stemming from the Indian cultures, the colors of Mexico tend to be bright and festive greens to reds or a pastel version of those colors.

With world boundaries blurring, and the cultures blending more than ever before, some of the strong color preferences and meanings are becoming less apparent. We now, for example, see many brides wearing traditional white dresses in all Asian countries.

Although the strict meanings of colors may be fading across countries, it is still a good practice to use colors carefully. We do not see brightly colored web sites or consumer products for the Japanese culture; instead, products are usually designed with pastel colors in mind.

If the same color palette is to be used across locales, it would be a good idea not to use primary colors in design because they may still carry negative or positive meanings in some cultures.

LAYOUT

Our final internationalization design consideration is layout. It is important not to assume anything about layout because all layout elements vary according to language (W3C, 2003). These elements include how a piece of text might be drawn on the screen, text alignment and justification, how much room translated text takes up, the direction it flows, line breaks, text wrapping, white spaces, word and letter spacing, text decorating (underline, overline, line-thru, text shadows, text underline solution for East Asian vertical writing and blinking), and where on the screen it should start.

Make sure there is enough room for text expansion. Layout has to accommodate different text properties and the change in context size due to translation. Therefore, it is important not to hardcode physical layout space, but to let it adjust dynamically. Use of dynamic layout managers is preferred over fixed-position layout because of the benefit of cross-platform automatic layout management for graphical user interface components.

SUMMARY

The purpose of this chapter was to list, in one place, a good sampling of the issues, guidelines, resources, and tips associated with international information display. This chapter should make it possible for the user interface practitioner to begin an internationalization project in a knowledgeable and productive way. The list of issues described in this chapter is by no means a complete list. But these issues will give cross-cultural designers a foundation on which experience can build a fuller understanding of lifestyles, preferences, customs, and work patterns.

REFERENCES

Allen, P. L., & Hall, J. (2003). *World currencies and abbreviations*. Retrieved January 22, 2004, from http://www.jhall.demon.co.uk/currency/

Aykin, N. (2003). *Internationalization and localization of Web sites*. Tutorial presented at the User Experience 2003 Conference.

GNOME Documentation Style Guide. (2003). *Chapter on writing for localization*. Published by: Sun Microsystems. Retrieved January 22, 2004, from http://developer.gnome.org/documents/style-guide/

Hewlett-Packard. (2003). *Printing with color—the meaning of color*. Retrieved January 22, 2004, from http://www.hp.com/sbso/productivity/color/meaning.html

Hoft, N. L. (1995). *International technical communication*. New York: Wiley.

Holzschlag, M. E. (1999). *Color my world. New architect: internet strategies for technology leaders*. Retrieved January 22, 2002, from http://www.webtechniques.com/archives/2000/09/desi/

Hongladarom, S. (2000). *Negotiating the global and the local: How Thai culture co-opts the Internet*. Retrieved January 22, 2004, from http://www.firstmonday.dk/issues/issue5_8/hongladarom/

ITU-T Recommendation E.164. (1997). *The International Public Telecommunication Numbering Plan*. International Telecommunications Union. Telecommunication Standardization Bureau. Geneva, Switzerland. Retrieved January 22, 2004, from http://www.itu.int/ITU-T/index.html

Japanese for Otakus. (2004). Retrieved January 22, 2004, from http://www.orange angel.com/otaku/april11.html

Jinsheng, M. (2004). *Colors and meanings in Chinese modern products*. Retrieved January 22, 2004, from http://www.aedo-to.com/eng/library/aedo-ba/aedo_02/art04.html

Kuhn, M. (1999). *Metric typographic units*. Retrieved January 22, 2004, from http://www.cl.cam.ac.uk/~mgk25/metric-type/html

Kuhn, M. (2001). *A summary of the international date and time notation*. Retrieved January 22, 2004, from http://www.cl.cam.ac.uk/~mgk25/iso-time.html.

Kuhn, M. (2003). *International standards paper size*. Retrieved January 22, 2004, from http://www.cl.cam.ac.uk/~mgk25/iso-paper.html.

Microsoft (Dr. International Team). (2002). *Developing international software*. Redmond, WA: Microsoft Press.

Morton, J. L. (2003). *Color matters*. Retrieved June 10, 2003, from www.colormattters.com

Morton, J. L. (2003). *Global color meanings*. Retrieved January 22, 2004, from http://www.colormatters.com/bubarc8-global.html

Prime, M., & Wilson, M. (2002). *World Wide Web Consortium (W3C) primer on the internationalisation and localisation of web pages*: W3C UK and Ireland Office 1 December 2002. Retrieved January 22, 2004, from http://www.w3c.rl.ac.uk/QH/QH_final_review_WP5/WD-int-primer.html

Russo, P., & Boor, S. (1993). How fluent is your interface? Designing for international users. INTERCHI'93. 342–347.

Spartan (1999). *Multimedia presentation on the colors of culture*. Retrieved January 22, 2004, from http://www.mastep.sjsu.edu/Alquist/workshop2/color_and_culture_files/frame.htm

Vekulenko, A. (2000). *Differences between point systems*. Retrieved January 22, 2004, from http://www.oberonplace.com/dtp/fonts/point.htm

Webjapanese.com. (2001). Retrieved June 10, 2003, from http://webjapanese.com/wj/ftp/dtp/images/kanji01_640_w.gif

Wired4success. (2003). *Color symbolism chart*. Retrieved January 22, 2004, from http://www.wired4success.com/colorsymbolism.htm

Chapter 3

User Interface Design and Culture

Aaron Marcus
Aaron Marcus and Associates, Berkeley, CA

INTRODUCTION

Over the past two and a half decades in the user interface (UI) design community, designers, analysts, educators, and theorists have identified and defined a somewhat stable set of UI components, that is, the essential entities and attributes of all UIs, no matter what the platform of hardware and software (including operating systems and networks), user groups, and contents (including vertical markets for products and services). That means these components can enable developers, researchers, and critics to compare and contrast UIs that are evidenced on terminals, workstations, desktop computers, Web sites, Web-based applications, information appliances, and mobile, wireless devices. (Marcus, 1998, 2001), among others, provides one way to describe these UI components, which is strongly oriented to communication theory and to applied theory of semiotics (Eco, 1976; Peirce, 1933). This philosophical perspective emphasizes communication as a fundamental characteristic of computing, one that includes perceptual, formal characteristics, and dynamic, behavioral aspects of how people interact through computer-mediated media. Picking up on Claude Lévi-Strauss's (2000) theory of human beings as sign makers and tool makers, my theory understands a UI as a form of dynamic, interactive visual literature as well as a suite of conceptual tools, and as such, a cultural artifact. The UI components are the following:

51

Metaphors

Metaphors are fundamental (Lakoff and Johnson, 1980) concepts communicated via words, images, sounds, and tactile experiences. Metaphors substitute for computer-related elements and help users understand, remember, and enjoy entities and relationships of computer-based communication systems. Metaphors can be overarching, or they can communicate specific aspects of UIs. An example of an overarching metaphor is the desktop metaphor to substitute for the computer's operating system, functions, and data. Examples of specific concepts are the trashcan, windows and their controls, pages, shopping carts, chatrooms, and blogs (Web logs). The pace of metaphor invention, including neologisms or verbal metaphor invention, is likely to increase because of rapid development and distribution through the Web and mobile devices of ever-changing products and services. Some researchers are predicting the end of the desktop metaphor era and the emergence of new fundamental metaphors (Gelerntner, 2000).

Mental Models

Mental models are structures or organizations of data, functions, tasks, roles, and people in groups at work or play. These are sometimes also called user models, cognitive models, and task models. Content, function, media, tool, role, goal, and task hierarchies are examples.

Navigation

Navigation involves movement through the mental models, that is, through content and tools. Examples of UI elements that facilitate such movement include those that enable dialogue, such as menus, windows, dialogue boxes, control panels, icons, and tool palettes.

Interaction

Interaction includes input/output techniques, status displays, and other feedback. Examples include the detailed behavior characteristics of keyboards, mice, pens, or microphones for input; the choices of visual display screens, loudspeakers, or headsets for output; and the use of drag-and-drop selection/action sequences.

Appearance

Appearance includes all essential perceptual attributes, that is, visual, auditory, and tactile characteristics,. Examples include choices of colors, fonts, animation style, verbal style (e.g., verbose/terse or informal/formal), sound cues, and vibration modes.

As computer-based communication has become a global enterprise, the development, marketing, and business communities have become aware of the impact of world cultures on UI design for global products and services. A clear need has emerged to determine optimum characteristics of suitably localized products and services based on market and user data to achieve both short-term and long-term success, without having to develop too many variations that might waste time and money in development, distribution, and maintenance.

Likewise, over many more decades, cultural anthropologists have identified and defined somewhat stable, fundamental analytical dimensions by which one may contrast and compare all world cultures. Culture, as evidenced at work, at home, in schools, and in families, may be defined in terms of a group's symbols, heroes, rituals, and values. In this chapter, culture refers, in general, to well-established groups whose characteristics a member may have significant difficulty to abandon, but may also include what might be called "temporary, voluntary life-style affinity groups" not recognized as cultures by some classical cultural anthropologists (Clausen, 2000).

For example, consider the international express-mail Web site page of DHL for shipping packages to Saudi Arabia shown in Fig. 3.1. Is this imagery appropriate for potential customers? This chapter seeks to provide initial resources for considering such questions. The issue is potent if one considers what DHL and FedEx, another international shipping company, were showing at the same point in time. Both DHL and FedEx had similar layouts, but the FedEx site showed several standard images, including an Asian woman with bare sleeves and uncovered face. This race, gender, and dress style is not typical for Saudi Arabia, while the DHL site showed an apparently Middle Eastern man in conservative Western business attire. The DHL page also showed a green logotype in Arabic script at the lower left, a typical reference via color and script to Saudi Arabian nationality and the Moslem religion.

As an example of a culture model, one may cite the work of the cultural analyst Geert Hofstede (Hofstede, 1997). He studied hundreds of IBM employees in 53 countries from 1978 to 1983 (see Table 3.1 for a display of his findings), from which study he developed five fundamental culture dimensions. These dimensions of culture are described in the following sections.

Power Distance

Power distance measures the extent to which people of a culture accept large or small distances of power in social hierarchies. For example, if an employee of a large organization within a culture typically has easy, informal access to the "boss," then one might assign a low power distance rating to that culture.

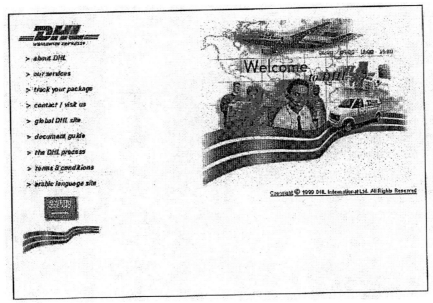

FIG. 3.1.   A DHL express-mail Web site page for Saudi Arabia, as displayed in 2000. Note that the Web site has been updated significantly since then.

Individualism Versus Collectivism

Individualism versus collectivism measures the orientation to individual or group achievements. If a culture typically focuses on group goals and orients promotion of achievements to the group, rather than to individuals of the group, then one might assign a high collectivist rating to that culture.

Masculinity Versus Femininity

Masculinity versus femininity measures the degree to which a culture does or does not separate traditional gender roles. Traditional gender roles assume that males are tough, task-oriented, individual- or group-oriented "warriors" going out in the world to conquer and defend, and females are tender, gentle homemakers and childrearers who are family- and people-oriented. Masculine cultures tend to support this traditional role separation; feminine cultures tend to diminish the differences, that is, to merge both roles.

Uncertainty Avoidance

Uncertainty avoidance measures the degree to which a culture is uncomfortable with uncertainty and seeks to reduce uncertainty, often in the pur-

TABLE 3.1

Hofstede's Cultural Indexes

	PDI		IDV		MAS		UAI		LTO	
	Rank	Score	Rank	Score	Rank	Score	Rank	Score	Rank	Score
Arab Countries	7	80	26/27	38	23	53	27	68		
Argentina	35/36	49	22/23	46	20/21	56	10/15	86		
Australia	41	36	2	90	16	61	37	51	15	31
Austria	53	11	18	55	2	79	24/25	70		
Bangladesh									11	40
Belgium	20	65	8	75	22	54	5/6	94		
Brazil	14	69	26/27	38	27	49	21/22	76	6	65
Canada	39	39	4/5	80	24	52	41/42	48	20	23
Chile	24/25	63	38	23	46	28	10/15	86		
China									1	118
Columbia	17	67	49	13	11/12	64	20	80		
Costa Rica	42/44	35	46	15	48/49	21	10/15	86		
Denmark	51	18	9	74	50	16	51	23		
East Africa	21/23	64	33/35	27	39	41	36	52		
Equador	8/9	78	52	8	13/14	63	28	67		
Finland	46	33	17	63	47	26	31/32	59		
France	15/16	68	10/11	71	35/36	43	10/15	86		
Germany FR	42/44	35	15	67	9/10	66	29	65	14	31
Great Britain	42/44	35	3	89	9/10	66	47/48	35	18	25
Greece	27/28	60	30	35	18/19	57	1	112		
Guatemala	2/3	95	53	6	43	37	3	101		
Hong Kong	15/16	68	37	25	18/19	57	49/50	29	2	96
India	10/11	77	21	48	20/21	56	45	40	7	61
Indonesia	8/9	78	47/48	14	30/31	46	41/42	48		
Iran	29/30	58	24	41	35/36	43	31/32	59		
Ireland (Rep of)	49	28	12	70	7/8	68	47/48	35		
Israel	52	13	19	54	29	47	19	81		
Italy	34	50	7	76	4/5	70	23	75		

(continued on next page)

TABLE 3.1 (continued)

	PDI		IDV		MAS		UAI		LTO	
	Rank	Score	Rank	Score	Rank	Score	Rank	Score	Rank	Score
Jamaica	37	45	25	39	7/8	68	52	13		
Japan	33	54	22/23	46	1	95	7	92	4	80
Malaysia	1	104	36	26	25/26	50	46	36		
Mexico	5/6	81	32	30	6	69	18	82		
Netherlands	40	38	4/5	80	51	14	35	53		
Philippines	4	94	31	32	11/12	64	44	44	21	19
Poland									13	32
Portugal	24/25	63	33/35	27	45	31	2	104		
Salvador	18/19	66	42	19	40	40	5/6	94		
Singapore	13	74	39/41	20	28	48	53	8	9	48
South Africa	35/36	49	16	65	13/14	63	39/40	49		
South Korea	27/28	60	43	18	41	39	16/17	85	5	75
Spain	31	57	20	51	37/38	42	10/15	86		
Sweden										
Sweden	47/48	31	10/11	71	53	5	49/50	29	12	33
Switzerland	45	34	14	68	4/5	70	33	58		
Taiwan	29/30	58	44	17	32/33	45	26	69	3	87
Thailand	21/23	64	39/41	20	44	34	30	64	8	56
Turkey	18/19	66	28	37	32/3	45	16/17	85		
Uruguay	26	61	29	36	42	38	4	100		
USA	38	40	1	91	15	62	43	46	17	29
Venezuela	5/6	81	50	12	3	73	21/22	76		
West Africa	10/11	77	39/41	20	30/31	46	34	54		
Yugoslavia	12	76	33/35	27	48/49	21	8	88		
Zimbabwe									19	25

Note. PDI = power distance index; IDV = individualism index; MAS = masculinity index; UAI = uncertainty avoidance index; LTO = long-term orientation index. Adapted from *Cultures and Organizations: Software of the Mind* by G. Hofstede, 1997, pp. 26, 55, 84, 113, 166. Original data copyright 1997 by Geert Hofstede.

suit of the truth. Cultures that focus on punctuality, formality, and explicit communication tend to rate high in uncertainty avoidance.

Long-Term Time Orientation

Long-term time orientation grows out of a long-term basis of some cultures in Confucian thought, which emphasizes patience. Naturally, this dimension seems strongest among Asian cultures steeped in Confucian philosophy. (For an example of communication influences and differences, see Oliver, 1971.)

An example of the interrelations of two cultural dimensions, power distance versus individualism-collectivism, for several countries appears in Figure 3.2.

De Mooij (2001) has argued for culture differences as a basis for differences of Internet and mobile behavior separate from economic factors. These culture dimensions, examples of culture bias in Web site designs, and the implications of culture dimensions on Web site and Web-based application designs are discussed at length elsewhere (Marcus, 2001; Marcus & Gould, 2000).

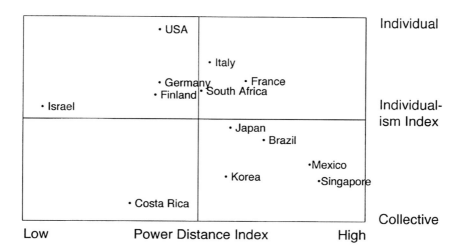

FIG. 3.2.   Chart of power distance versus individualism-collectivism for selected countries. Adapted from *Cultures and Organizations: Software of the Mind* by G. Hofstede, 1997, p. 54. Copyright 1997 by McGraw-Hill. Adapted with permission of McGraw-Hill.

MAPPING DIMENSIONS TO COMPONENTS

Although these definitions of dimensions and components are not universally accepted, they are stable enough to consider a next phase of understanding the relationship of culture to UIs: mapping components of UIs to
dimensions of culture, and vice versa. Being able to understand the relationships of UI components to a particular culture dimension is a step forward to
mapping particular UI designs to one or more cultures. Reliably being able
to understand and predict the use, acceptance, and enjoyment of particular
UIs marks significant progress in developing consistently successful products
and services. In addition, with greater understanding of the relationships, it
becomes more feasible to develop databases, such as databases of culture
phenomena or catalogs of UI design patterns (Tidwell, 1999) in which culture dimensions play a more significant role in the attributes and tools.

Such tools might enable designers to eventually "semiautomatically" adjust UI designs per culture dimension, to sets of countries with similar-cul-
ture characteristics (e.g., adjusting UI designs to be suitable for all countries
that were once part of the British Empire), and ultimately, per culture. These
"culture-design" tools someday may be ubiquitous and conventional, similar
to the color adjustment tools today of a graphics product like Adobe
Photoshop™, which permits a designer to choose color sets that are safe for
the Web. A future scenario might find designers (from one or more cultures)
designing a UI for a specific culture, then asking the computer to adjust the
entire UI design for another culture, say Culture X, to make it "Culture
X-safe." This design might be analyzed by the designers or submitted to users
of the target culture for feedback. In an interactive process, the design might
be adjusted, including the adjustment of culture parameters to reflect ongoing changes in the target cultures, which are not themselves static.

To explore possible relations of culture dimensions to UI components, I,
sometimes in conjunction with Emilie Gould of Rensselaer Polytechnic Institute (goulde@rpi.edu), have been giving tutorials about cross-cultural
communication since 1999 in several countries, including North America,
Europe, South Africa, and Japan. In the tutorials, we ask participants to initiate analysis and discussion of the relation of culture dimensions to UI
components. The conceptual challenge is to fill in a matrix like that of Fig.
3.3. To stimulate further study, an initial mapping appears here. This mapping, in greater detail, with the appropriate conditional influences, would
be the basis for detailed heuristics, design pattern catalogs, and, eventually,
a kind of artificial-intelligence-based application that could assist in designing UIs to be culturally conservative or radical according to the needs/de-
sires of those doing the designing.

	Metaphors	Mental Models	Navigation	Interaction	Appearance
Power Distance					
Individualism-Collectivism					
Masculinity-Femininity					
Uncertainty Avoidance					
Long-term Time Orientation					

FIG. 3.3. Matrix mapping culture dimensions to user interface components.

In the analysis that follows, for a given culture dimension, each subsection provides characteristic examples, as determined by personal experience, not yet by detailed study, of UI elements listed per UI component, as described previously. Although this is an initial categorization by assertion, I believe it nevertheless has pedagogic and practical value.

Power Distance

Metaphors

High: Government or corporate institutions and buildings; objects with clear hierarchies, e.g., the human body, schools, government, monuments, etc.

Low: Informal, or popular institutions, buildings, and objects, that emphasize equality, options; Montessori schools, play/games, public spaces, etc.

Mental Models

High: Complex, highly organized, highly categorized, highly populated structures, e.g., large corporate and/or government organizational models or charts; reference data with little or no relevancy ranking

Low: Simple, informally organized and categorized structures; less structured data with some or much relevancy

Navigation

High: Restricted access, choices; authentication; passwords; prescribed routes
Low: Open access, multiple options, sharable paths

Interaction

High: Severe error messages: "Entry Forbidden," "You are wrong;" wizards or guides lead usage
Low: Supportive error messages, cue cards; many user-driven options available

Appearance

High: Images of leaders; national/corporate/government themes, slogans, insignia, logos, symbols, typefaces, layouts, colors; official music, anthems; formal speech
Low: Images of people, groups; daily activities; popular music, symbols, typefaces, layouts, colors; informal speech

Individualism Versus Collectivism

Metaphors

Individualist: Action-oriented, tool-oriented
Collectivist: Relationship-oriented, content-oriented

Mental Models

Individualist: Product- or task-oriented
Collectivist: Role-oriented

Navigation

Individualist: Individual paths; popular choices, celebrity choices; stable across roles; customizable
Collectivist: Group-oriented, official choices; changes per role

Interaction

Individualist: Keyword searches; active-oriented; multiple devices; customizable;
Collectivist: Limited, official devices; role-driven

Appearance

Individualist: Images of products, people; low context; hyperbolic, dynamic speech; market-driven topics, imagery, language; customizable; direct, active verbs
Collectivist: Images of groups, organizations; images of roles; high context; official, static terminology; institution-driven topics, imagery, language; passive verbs

Masculinity Versus Femininity

Note: This subsection's contents are differently organized and may be misinterpreted without considering the fact that the examples list items for *traditional gender roles*, not for masculine cultures or feminine cultures *per se*. Masculine cultures would maintain clear separation of such examples; feminine cultures would tend to combine or merge them, making fewer gender-oriented distinctions.

Metaphors

Masculine: Sports-oriented; competition-oriented; work-oriented
Feminine: Shopping carts; family-oriented, people-oriented

Mental Models

Masculine: Work/business structures; high-level, "executive views;" goal-oriented
Feminine: Social structures; detailed views; relationship-oriented

Navigation

Masculine: Limited choices, synchronic
Feminine: Multiple choices; multitasking, polychronic

Interaction

Masculine: Competitive-game-oriented; mastery-oriented; individual-oriented
Feminine: Practical, function-oriented; cooperation-oriented; team oriented

Appearance

Masculine: "Masculine" colors, shapes, sounds
Feminine: "Feminine" colors, shapes, sounds; acceptance of cuteness

Uncertainty Avoidance

Metaphors

High: Familiar, stable, clear references to daily life; representation
Low: Novel, unusual references; abstraction

Mental Models

High: Simple, explicit, clear articulation; limited choices; binary logic
Low: Tolerance for ambiguity, implicit structures or relations, complexity; fuzzy logic, i.e., multivariate, non-binary logic.

Navigation

High: Desire for limited, clear organized options; tolerance for complex, fine tuning controls to "master" or "control" a situation, e.g., advanced searches on the Web, consumer electronics controls.
Low: Tolerance for ambiguous, possibly redundant options; tolerance for risk, gambling; tolerance for simple controls, e.g., simple searches on the Web, or www.Google.com's "I Feel Lucky" button.

Interaction

High: Precise, complete, detailed input and feedback of status; devices that permit finetuning.
Low: General, limited, or ambiguous input and feedback of status; devices that may have gross tuning.

Appearance

High: Simple, clear, consistent imagery, terminology, sounds; highly redundant coding of perceptual cues.

Low: More varied, ambiguous, less-consistent imagery, terminology, sounds; tolerance for more perceptual characteristics involved in purely ornamental or aesthetic use; less redundant coding of perceptual cues.

Long-Term Time Orientation

Metaphors

Long: Stable family references; paternalistic imagery and references: Father, Mother; stable institution/organization references, e.g., universities, major businesses, the Mafia, Chinese state businesses, IBM in 1950s, etc.; concrete metaphors

Short: Interchangeable roles, jobs, objects; less concrete, more abstract metaphors

Mental Models

Long: Love/devotion; social coherence, responsibility, support
Short: Liberty: social incoherence, social irresponsibility, efficiency

Navigation

Long: Tolerance for long paths, ambiguity; contemplation-oriented, reliance on personal advisor; context-oriented

Short: Bread-crumb trails, clear taxonomies; quick-results; logical inference-oriented, function or action-oriented

Interaction

Long: Preference for face-to-face communication, harmony; personalized messages; more links to people; live chats; interaction as "asking"

Short: Distance communication accepted as more efficient; anonymous messages tolerated; conflict tolerated, even encouraged; performance-critical communication

Appearance

Long: Cultural markers: flags, colors, national images; soft focus; warm, fuzzy images; pictures of groups inviting participation, suggestions of intimacy and close social distance

Short: Minimal and focused images; short borders, lines, edges; concentration on showing task or product

This initial taxonomy represents an attempt to group UI attributes in a manner that highlights the connection to culture dimensions. A more definitive mapping clearly requires extensive effort, but this initial construct points the way.

FUTURE STUDY OF CULTURE IN RELATION TO USER INTERFACE DESIGN

The preceding analysis is not the only way to relate UI design and culture. Recent publications highlight other cultural issues that also should be considered in understanding how cultural differences might impact UI design. These other reference dimensions could/should be mapped to culture dimensions and to UI components in future research.

First of all, Clausen (2000) argued that classical culture in the United States no longer exists, at least as defined by cultural anthropologists in the past. Culture in earlier circumstances represented group environments that were difficult to escape. Now, many uses of the term *culture* refer simply to matters of choice, not requirement, in other words, to lifestyles. The strict use of the term would not be applied to, for example, consumer product affinity groups, such as owners of Volvo automobiles. Consequently, designers may need to debate how strictly they apply the concepts of traditional anthropology to current product design and to performance, preference, and enjoyment differences among target groups. As noted earlier, this chapter adopts a looser definition of culture, rather than a stricter one, for practical purposes in UI design.

Some research about how different cultures use different UI constructs has appeared. Choong and Salvendy (1998), for example, studied the use of icons in Chinese UIs. For their article "Implications for Design of Computer Interfaces for Chinese Users in Mainland China," Choong and Salvendy (1999) investigated the impact of cultural differences on computer performance for 40 Chinese subjects residing in mainland China and 40 U.S. users and the design of appropriate UIs for the Chinese users. Their results are also commented on by Carroll (1999).

Choong and Salvendy found, in general, that U.S. users rely much more on logical inference and categorization. These users tend to classify by functions, analyze components, and infer common features. In contrast, the Chinese users relied more on relations and contexts. These users tended to classify by interdependence within wholes, to rely on subjective experience without sharp differences between oneself and others or between facts and concepts.

They investigated sorting the contents of one's house and found contrasting sorting styles. The Chinese tend to sort by thematic groupings whereas the U.S. users tend to sort by functions. They found that when users were given sorting tasks and the culturally appropriate sorting methods, each had the lowest error rates, and in contrast with the wrong kind of tool, high error rates. Furthermore, test results of Chinese users using their preferred thematic structures show better memory performance, speed, and accuracy. The Chinese also performed better using concrete metaphors. These investigations are in general accordance with the heuristic analysis provided earlier and seem to have strong implications for significant performance differences of users given culturally inappropriate UI components.

Robert Cialdini, professor at the University of New Mexico, has written about the dimensions of persuasion (Cialdini, 2001) by which people convince others to think or act in a particular way. He identifies the following dimensions of persuasion, that is, factors that might persuade one to follow another's request:

- Authority: One might act out of respect or fear of another's authority, power, or position.
- Consistency: One might act out of belief in the consistent following of rules of legal or sanctioned behavior under the conditions of the request and appropriate rules of behavior.
- Liking: One might act out of consideration of positive personal feelings toward the person making the request.
- Reciprocation: One might act in consideration of one's own requests in the past or desired reciprocity when making future requests.
- Scarcity: One might act out of perceived scarcity of referred-to objects, persons, and the perceived benefits of following the request.
- Social validation: One might act out of consideration of value, status, or other social benefit accrued by following the request.

Although several of these dimensions combine in any situation to lead to someone's actions, Cialdini explains that different cultures typically emphasize one factor over others. For example, he comments that an American might decide to honor a request from another person in an office based on whether that person had recently done him or her a favor, whereas a German would consider the rules regarding the request, a Spaniard would consider whether the person is a friend, and a Chinese office worker would consider the person's authority to issue the request. Although Cialdini does not consider Web site persuasion and does not elaborate extensively on cultural differences that might affect e-commerce and m-commerce, his analysis clearly opens the door to such considerations.

As an example of additional dimensions in relation to UI design and culture, consider Bailey, Gurak, and Konstan (2001), who developed a set of dimensions for establishing trust on the Web. They assert that the following are the key dimensions:

- Attraction: Attractive people are trusted more.
- Dynamism: Greater activity builds trust (e.g., moving hands, text).
- Expertness: Relevant skills are trusted.
- Faith: Belief in a predictable future builds trust.
- Intentions: Revealed objectives and goals are presumed to be honest and are trusted.
- Localness: Local personalities, sites, terminology, references are presumed to have similar values and behavior; hence, they are trusted more.
- Reliability: Dependable, predictable, consistent behavior or outcomes promote trust.

Bailey (personal communication, June, 2001) commented that he feels these dimensions vary in their importance among different cultures globally, and need to be examined in the light of these differences in order to predict the influence of each within varying cultures.

In *Frames of Mind, The Theory of Multiple Intelligences* (1985), Gardner presents his theory of the dimensions of intelligence, considers much education and communication biased to verbal comprehension, and urges greater diversity in communication, evaluation, and learning. His dimensions are the following:

- Verbal/Image comprehension.
- Word/image fluency.
- Numerical/graphical fluency.
- Spatial visualization.
- Associative memory.
- Perceptual speed.
- Reasoning.
- Image: Self/Other awareness.

It seems reasonable that different cultures might place different emphases on which dimensions are valued within that culture and emphasized in education, at work, and at home. A future task for UI design theory and practice is to explore how UI components, and specific patterns of combinations, would relate to these dimensions of intelligence and to the varying roles given to them by different cultures.

Richard Nisbett, a social psychologist from the University of Michigan, has studied how culture molds habits of thought. He uses examples of the differences in descriptions of what Japanese and North Americans observe in describing simple scenes, for example, what happens in a fish tank (Nisbett, 2001, p. 6). The Japanese observers note relations among the fish, the aquatic flora, and the simple environment. North American observers, on the other hand, tend to focus only on the actions and attributes of the primary fish. The contrast is stark: One culture is much more oriented to figure-field relationships; the other culture concentrates on the figure. Similar differences have been noted by Kaiping Peng at the University of California at Berkeley. Peng (2000) noted that Chinese students seem less eager to resolve contradictions than U.S. students. Cultural factors seem to run deeply in mental processes. One possible conclusion is that Western scholars may be in error to assume that there is, or should be, a universal focus on reasoning, categorization, and linear cause-and-effect explanations of situations and events. Such conclusions differ markedly from classical assumptions that would affect the design of UIs.

David Landes, professor of history at Harvard University, argues in "Culture Makes Almost All the Difference" (2000) that social attitudes are more important than politics and economics in determining why some societies are richer than others. This version of the thesis that culture matters greatly is at odds with assumptions of market economists and liberal and Marxist philosophers, who believe that political and economic factors are of primary importance. Hofstede (1997) has a more mixed approach. He recognizes that both cultural as well as political/economic factors determine whether certain cultures are dominant at different points in history.

Hofstede comments that cultural orientations are extremely deeply embedded in cultures over hundreds and thousands of years; consequently, he feels that even modern communication media have not dislodged these cultural orientations. In fact, cultures tend sometimes to concentrate harder on preserving their approaches given the encroachment of alien cultures. On the other hand, recent news events seem to argue otherwise. For example, a recent article comments that Japan's 10-year economic decline has stimulated the rise of individualism (Ono & Spindle, 2000).

CONCLUSIONS

The design of products and services for the Web and mobile devices fosters the need for good cross-cultural communication in UI design. English-speaking countries constitute 8% of the world's population, but by

2005, approximately 75% of Internet users will be non-English speaking. It seems likely that cultural factors will need to be considered more frequently. Already, 80% of corporate Web sites in Europe offer more than English even though launching multilanguage Web site portals with 11 European languages is a significant burden to operations.

Consequently, cross-cultural analysis and design issues need to be considered more integrally in planning stages, and developers need checklists and guidelines to assist them in their design phases. Having a better understanding of the mappings of culture dimensions to UI components, as well as to such dimensions as trust or intelligence, may inform designers to make better decisions about usability, aesthetics, and emotional experience. Ultimately, computer-assisted tools may result. As indicated earlier, increased performance, preference, and enjoyment seem likely outcomes of culturally appropriate UI design.

ACKNOWLEDGMENTS

I acknowledge the assistance of Emilie Gould, cited earlier, in organizing and conducting the tutorials, and the tutorial participants, through which portions of the material in this chapter were developed. I also acknowledge two works of mine (Marcus, 2001, 2002), edited portions of which appear in this chapter.

REFERENCES

Bailey, B. P., Gurak, L. J., & Konstan, J. A. (2001). *An examination of trust production in computer-mediated exchange.* Paper presented at the seventh Human Factors and the Web 2001.

Cialdini, R. (2001, February). The science of persuasion. *Scientific American, 284*(2), 76–81.

De Mooij, M. (2001). Internet and culture. Paper presented at. *Internet, Economic Growth and Globalization* organized by Institute for International and Regional Economic Relations, Gerhard-Mercator University, Duisburg, Germany, 8 August 2001.

Carroll, J. M. (1999). Using design rational to manage culture-bound metaphors for international user interfaces. *Proceedings, International Workshop on International Products and Services*, Rochester, NY, pp. 125–131.

Choong Y., & Salvendy, G. (1998). Designs of icons for use by Chinese in mainland China. In Interacting with computers. *International Journal of Human–Computer Interaction*, 9:4, February 1998, pp. 417–430. Amsterdam: Elsevier.

Choong, Y., & Salvendy, G. (1999). Implications for design of computer interfaces for Chinese users in Mainland China. *The International Journal of Human-Computer Interaction*, 11:1 1999, pp. 29-46. Amsterdam: Elsevier. Cialdini, Robert.

(2001). The Science of Persuasion, *Scientific American*, Vol, 284, No. 2, February 2001, pp. 76–81 (www.influenceatwork.com).

Clausen, C. (2000). *Faded mosaic: The emergence of post-cultural America*, Ivan R. Dee.

Eco, U. (1976). *A theory of semiotics*. Bloomington: Indiana University Press.

Gardner, H. (1985). *Frames of mind: The theory of multiple intelligences*. New York: Basic Books.

Gelerntner, D. (2000). *The second coming: A manifesto*. http://www.edge.org/documents/archive/edge70.html

Hofstede, G. (1997). *Cultures and organizations: Software of the mind*. New York: McGraw-Hill.

Lakoff, G., & Johnson, M. (1980). *Metaphors we live by*. Chicago: University of Chicago Press.

Landes, D. (2000). Culture makes all the difference. In L. E. Harrison & S. P. Huntington (Eds.), *Culture matters: How values shape human progress* (pp. 2–13). New York: HarperCollins.

Lévi-Strauss, C. (2000). *Structural anthropology*. Trans. Claire Jacobson and Brooke Schoepf. New York: Basic Books.

Marcus, A. (1998). Metaphors in user-interface design. *Journal of Computer Documentation, 22*(2), 43–57).

Marcus, A. (2001). Cross-cultural user-interface design. In M. J. Smith & G. Salvendy (Eds.), *Proceedings of the Human–Computer Interface International Conference* (Vol. 2, pp. 502–505). Mahwah, NJ: Lawrence Erlbaum Associates.

Marcus, A., & Gould, E. W. (2000). Crosscurrents: Cultural dimensions and global web user-interface design. *Interactions, (7)*4, 32–46.

Nisbett, Richard E., Kaipeng Peng, Incheol Choi, & Ara Norenzayan (2001). Culture and systems of thought: Holistic vs. analytical cognition, *Psychological Review,* 108, 291–310.

Oliver, R. T. (1971). *Communication and culture in ancient India and China*. Syracuse, NY: Syracuse University Press.

Ono, Y., & Spindle, B. (2000, December 30). Japan's long decline makes one thing rise: Individualism. *Wall Street Journal*, p. A1.

Peirce, C. S. (1933). Existential graphs. In C. Hartshorne & P. Weiss (Eds.), *Collected papers of Charles Sanders Peirce* (Vol. 4, pp. 293–470). Cambridge, MA: Harvard University Press.

Peng, K. (2000). *Readings in cultural psychology: Theoretical, methodological and empirical developments during the past decade (1989–1999)*. New York: Wiley.

Tidwell. J. (1999). *Common ground: A pattern language for human–computer interface design*. Retrieved February 20, 2002, from www.mit.edu/~jtidwell/interaction_patterns.html

Additional Useful References

Atchison, Jean. (1996). *The atlas of languages: The origin and development of languages throughout the world*, Facts On File. New York, 1966.

Alvarez, G. M., Kasday, L. R., & Todd, S. (1998). How We Made the Web Site International and Accessible: A Case Study. *Proceedings* (CD-ROM), *Fourth Human Factors and the Web Conference*, Holmdel, New Jersey.

American Institute of Graphic Arts (AIGA). (1981). *Symbol signs*. Visual Communication Books. New York: Hastings House.

Associated Press. (2001). Saudi Arabia issues edict against Pokemon, *San Francisco Chronicle*, 27 March 2001, p. F2.

Barber, W., & Badre, A. (1998). Culturability: The merging of culture and usability. *Proceedings of the Fourth Human Factors and the Web Conference*.

Bliss, C. K. (1995). *Semantography*. Sidney, Australia: Semantography Publications.

Campbell, George L. (2000). *Concise compendium of the world's languages*. Sandpoint, ID: MultiLingual Computing and Technology.

Comrie, Bernard, Stephen Matthews, and Maria Polinsky (Eds.). (2000). *The Atlas of languages*. Sandpoint, ID: MultiLingual Computing and Technology.

Coriolis Group. (1998). *How to build a successful international web site*. Scottsdale, AZ: Coriolis Group.

Cox, Jr. Taylor (1994). *Cultural diversity in organizations*, San Francisco: Berrett-Koehler Publishers.

Crystal, David. (1987). *The Cambridge Encyclopedia of Language*. Cambridge: Cambridge University Press.

Daniels, Peter T., & William Bright. (Eds.). (2000). *The World's Writing Systems*. Sandpoint, ID: MultiLingual Computing and Technology.

Day, Donald L., Elisa M. del Galdo, and Girish V. Prabhu. (Eds.). (2000). Designing for Global Markets 2, *Second International Workshop on Internationalisation of Products and Systems*, 13–15 July, Baltimore, MD. Rochester, NY: Backhouse Press.

Day, Donald. (2000). Gauging the Extent of Internationalization Activities. *In the Proceedings of the Second International Workshop on Internationalization of Products and Systems*, 13–15 July 2000, Baltimore, MD, pp. 124–136. Rochester, NY: Backhouse Press.

DelGaldo, E., & Nielsen, J. (Eds.) (1996). *International user interfaces*. New York: John Wiley and Sons.

Doi, Takeo. (1973). *The anatomy of dependence*. New York: Kodansha International.

Doi, Takeo. (1986). *The anatomy of self: The individual versus society*. New York: Kodansha International.

Dreyfuss, Henry. (1966). *Symbol sourcebook*. New York: Van Nostrand Rhinehold.

Elashmawi, Farid, and Harris, Philip R. (1998). *Multicultural management 2000: Essential cultural insights for global business success*. Houston: Gulf Publishing Co.

Fernandes, T. (1995). *Global interface design: A guide to designing international user interfaces*. Boston: AP Professional.

French, Tim, and Smith, Andy. Semiotically enhanced web interfaces: Can semiotics help meet the challenge of cross-cultural design? *In the Proceedings of the Second International Workshop on Internationalization of Products and Systems*, 13–15 July, Baltimore, MD, pp. 23–38. Rochester, NY: Backhouse Press.

Frutiger, A. (1997). *Signs and symbols: Their design and meaning*. Cham, Switzerland: Syndor Press GmbH.

Goode, Erica. (2000). How culture molds habits of thought. *New York Times*, 8 August 2000, p. D1 ff.

Gould, Emilie. W., Zakaria, Norhayati, and Shafiz Affendi Mohd. Yusof. (2000). Applying culture to website design: A comparison of Malaysian and US websites. In the *Proceedings of the IPCC/SIGDOC Conference*, Boston, 25 September 2000.

Graham, Tony. (2000). Unicode™: A Primer. Sandpoint, ID: MultiLingual Computing and Technology.

Hall, E. (1969). *The hidden dimension*. New York: Doubleday.

Harel, Dan, and Girish Prabhu. (1999). Global User Experience (GLUE), Design for cultural diversity: Japan, China, and India. *Proceedings of the First International Workshop on Internationalization of Products and Systems*, 20–22 May 1999, pp. 205–216. Rochester, NY: Backhouse Press.

Harris, John, and McCormack, Ryan. (2000). Translation is Not Enough: Considerations for Global Internet Development. San Francisco: Sapient. Retrieved February 18, 2002, from http://www.sapient.com/pdfs/strategic_viewpoints/sapient_globalization.pdf

Harris, Philip R., and Moran, Robert T. (1993). *Managing cultural differences: High performance strategies for a new world of business*. Houston, TX: Gulf Publishing Co.

Helander, Martin G., et al. (Eds.). (1997). *Handbook of human–computer interaction*. Amsterdam, Netherlands: Elsevier Science.

Hendrix, Anastasia. (2001). The nuance of language. *San Francisco Chronicle*, 15 April, 2001, p. A10.

Herskovitz, Jon. (2000). J-Pop Takes Off: Japanese music, movies, TV shows enthrall Asian nations. *San Francisco Chronicle*, 26 December 2000, p. C2.

Hoft, Nancy L. (1995). *International technical communication: How to export information about high technology*. New York: John Wiley and Sons, Inc.

Honold, Pia. (1999). Learning how to use a cellular phone: Comparison between German and Chinese users. *Journal of Society of Technical Communication, 46*(2), May 1999, pp. 196–205.

International Organization for Standardization (ISO). (1996). *ISO 7001: Public Information Symbols of The American National Standards Institute*. Geneva: ISO.

International Organization for Standardization (ISO). (1993). *ISO 7001: Public Information Symbols: Amendment 1 of The American National Standards Institute*. Geneva: ISO.

International Standards Organization (ISO). (1989). *Computer Display Color: Draft Standard Document 9241-8*, Geneva: ISO.

Iwayama, M., Tokunaga, T., & Tanaka, H. (1990). A method of calculating the measure of salience in understanding metaphors. *In the Proceedings of the American Association for Artificial Intelligence (AAAI)*, 1990, pp. 298–303.

Kimura, Doreen. (2002). Sex differences in the brain. *Scientific American, 267*(3), pp. 118–125.

Konkka, Katja, and Koppinen, Anne. (2000). Mobile devices: Exploring cultural differnces in separating professional and personal time. In the *Proceedings of the Second International Workshop on Internationalization of Products and Systems*, 13–15 July, Baltimore, MD, pp. 89–104. Rochester, NY: Backhouse Press.

Kurosu, M. (1997), Dilemma of usability engineering. In Salvendy, G., Smith, M., and Koubek, R. (Eds.): *Design of Computing Systems: Social and Ergonomics Considerations* (Vol. 2), Proceedings of the Seventh International Conference on Human–Computer Interaction. San Francisco, USA, August 24–29, 1997, pp. 555–558. Amsterdam: Elsevier.

Lanham, R. A. (1991). *A handlist of rhetorical terms* (2nd ed.). Berkeley: University of California Press.

Leventhal, L., et al. (1996). Assessing user interfaces for diverse user groups: Evaluation strategies and defining characteristics. *Behaviour and Information Technology, 15*(3), pp. 127–139.

Lewis, Richard. (1991). *When cultures collide*. London: Nicholas Brealey.

Lingo Systems. (1999). *The guide to translation and localization*. Los Alamitos, CA: IEEE Computer Society.

The Localization Industry Standards Association (USA). (1999). *The localization industry primer*. Féchy, Switzerland: LISA.

Marcus, Aaron. (1983). Visual rhetoric in a pictographic-ideographic narrative. In the *Proceedings of the Second Congress of the International Association for Semiotic Studies*, Vienna, July 1979. Tasso Borbé (Ed.), *3*(6), pp. 1501–1508. Berlin: Mouton Publishers,

Marcus, Aaron. (1992). *Graphic design for electronic documents and user interfaces*. Reading: Addison-Wesley.

Marcus, Aaron. (1993). Human communication issues in advanced UIs. *Communications of the ACM, 36* (4), pp. 101–109.

Marcus, Aaron. (1993). Designing for diversity. In the *Proceedings of the 37th Human Factors and Ergonomics Society*. Seattle, Washington, 11–15 October 1993, Vol. 1, pp. 258–261.

Marcus, Aaron. (1995). Principles of effective visual communication for graphical user-interface design. In Baecker, Grudin, Buxton and Greenberg (Eds.), *Readings in Human–Computer Interaction* (2nd ed., pp. 425–441). Palo Alto, CA: Morgan Kaufman.

Marcus, A. (2000). International and intercultural user interfaces, in *User interfaces for all*, ed. Dr. Constantine Stephanidis, Lawrence Erlbaum Associates Publishers, New York, 2000, pp. 47–63.

Marcus, Aaron, and Chen, Eugene. (2002). Designing the PDA of the future. *Interactions, 9*:1, pp. 32–44.

Marcus, Aaron (2002). User-interface design and culture dimensions, *Workshop on Internationalization*, Proceedings. CHI 2002, Minneapolis, MN, 21-24 April 2002. in press.

Marcus, Aaron, et al. (1999). Globalization of user-interface design for the web, Proc., 5th *Conference on Human Factors and the Web*, 3 June 1999, NIST, Gathersburg, MD, avail. from www.tri.sbc.com/hfweb.

McLuhan, M. (1964). Understanding media: The extensions of man. New York: McGraw-Hill.

Neale, D. C. and John M. C. (1997). The role of metaphors in user-interface design. In Helander, M., Landauer, T. K. and Prabhu, P. (Eds.). *Handbook of human–computer interaction* (2nd ed), pp. 441–462. Amsterdam: Elsevier Science.

Neustupmy, J. V. (1987). Communicating with the Japanese. Tokyo: *The Japan Times*.

Nielsen, J. (Ed.) (1990). *Designing user interfaces for international use* (Vol. 13). Advances in human factors/ergonomics, Amsterdam: Elsevier Science.

Nisbett, R. E. (2003). *The geography of thought: How Asians and Westerners think differently … and why*. New York: Free Press.

Lee, Ook, The Role of cultural protocol in media choice in a confucian virtual workplace, *IEEE Transactions on Prof. Comm.*, Vol. 43, No. 2, June 2000, pp. 196–200.

Olgyay, N. (1995). *Safety symbols art*. New York: Van Nostrand Reinhold.

Ota, Yukio. (1987). *Locos: Lovers communications system* (in Japanese), Pictorial Institute, Tokyo, 1973. The author presents the system of universal writing that he invented.

Ota, Yukio, *Pictogram design*. Tokyo: Kashiwashobo.

Perlman, Gary. (2000). ACM SIGCHI Intercultural Issues. In the *Proceedings of the Second International Workshop on Internationalization of Products and Sysems*, 13–15 July, Baltimore, MD, pp. 183–195. Rochester, NY: Backhouse Press.

Pierce, T. (1996). *The international pictograms standard*. Cincinnati, OH: ST Publications.

Pirkl, James J. (1994). *Transgenerational design: Products for an aging population*. New York: Van Nostrand Reinhold.

Prabhu, G. V., Chen, B., Bubie, W., and Koch, C. (1997). Internationalization and localization for cultural diversity. In Salvendy, G., Smith, M., & Koubek, R. (Eds.): *Design of Computing Systems: Cognitive Considerations (Vol. 1)*, Proceedings of the Seventh International Conference on Human–Computer Interaction, San Francisco, USA, August 24–29, 1997, pp. 149–152. Amsterdam: Elsevier.

Prabhu, Girish and Dan Harel. (1999). GUI Design preference validation for Japan and China: A Case for KANSEI Engineering? In the *Proceedings of the Eighth International Conference on Human–Computer Interaction*, Munich, Germany 22–26 August 1999, pp. 521–525. Amsterdam: Elsevier.

Prabhu, Girish V., and delGaldo, Elisa M. (Eds.). (1999). Designing for global markets 1, Proceedings of the *First International Workshop on Internationalization of Products and Systems*, 20–22 May 1999, Rochester, NY. New York: Backhouse Press.

Rowland, Diana. (1985). *Japanese business etiquette*. New York: Warner Books.

Shahar, Lucy, and Kurz, David. (1995). *Border crossings: American interactions with Israelis*. Yarmouth, Maine: Intercultural Press.

Sheppard, Charles, and Jean Scholtz (1999). The effects of cultural markers on website use. In the Proceedings of the *Fifth Conference on Human Factors and the Web*.

Stephanidis, C., et al. (Eds.). (1998). Toward an information society for all: An International R&D agenda. *International Journal of Human–Computer Interaction*, 10(2), pp. 107–134.

Stephanidis, Constantine. (Ed) (2000). *User interfaces for all*. New York: Lawrence Erlbaum Associates Publishers.

Stille, Alexander. (2001). An old key to why countries get rich: It's the culture that matters, some argue anew. *New York Times*, 13 January 2001.

Storti, Craig. (1994). *Cross-cultural dialogues: 74 brief encounters with cultural difference*. Yarmouth, ME: Intercultural Press.

Tannen, Deborah. (1986). *That's not what I meant!* New York: Ballantine Books.
Tannen, Deborah. (1990). *You just don't understand: Women and men in conversation.* New York: William Morrow and Company.
Trompenaars, F., & Hampden-Turner, C. (1998). *Riding the waves of culture.* New York: McGraw-Hill.
Vickers, Ben. (2000). Firms push to get multilingual on the web. *Wall Street Journal,* 22 November 2000, p. B11A.
Victor, David A. (1992). *International business communication.* New York: HarperCollins Publishers.
Waters, Crystal. (1997). *Universal web design: Reach more viewers, improve site appearance, and enhance your marketability through efficient design.* Indianapolis, Indiana: New Riders Publishing.
Yardley, Joanathan. (2000). Faded mosaic nixes idea of 'cultures' in U.S. *San Francisco Examiner,* 7 August 2000, p. B3.
Yeo, Alvin Wee. (2001). Global-software development lifecycle: An exploratory study. In the *Proceedings of Special Internet Group for Computer–Human Interaction (SIGCHI),* 31 March–5 April 2001, Seattle, WA, pp. 104–111.

URLs and Other Resources

Selected URLs and other resources from a list maintained by the author and available by contacting him at <Aaron.Marcus@AMandA.com> are the following:

ACM/SIGCHI Intercultural Issues database: www.acm.org/sigchi/intercultural/
ACM/SIGCHI Intercultural listserve: chi-intercultural@acm.org. Moderator: Donald Day <d.day@acm.org.>.
African-American Websites: www.bet.com, www.netnoir.com, www.blackfamilies.com
Bibliography of Intercultural publications: www.HCIBib.org//SIGCHI/Intercultural
Color: www.colortool.com
Cultural comparisons: www.culturebank.com
Culture resources: www.webofculture.com, www.acm.org/sigchi/intercultural/
Digital divide: www.digitaldivide.gov, www.digitaldivide.org, www.digitaldividenetwork.org
Global business: www.globalbusinessmag.com
Glossary, six languages: www.bowneglobal.com/bowne.asp?page=9&language=1
Indian culture: www.indiagov.org/culture/overview.htm
Internationalization providers: Basis Technology <www.basistech.com>, CIJ America <www.cij.com>, www.Logisoft.com
Internationalization resources: www.world-ready.com/r_intl.htm, www.worldready.com/biblio.htm
Internet statistics by language: www.euromktg.com/globstats/index.html, www.worldready.com/biblio.htm
Java Internationalization: http://java.sun.com/docs/books/tutori
Localization: www.lisa.org/home_sigs.html

Localization providers (selected): www.Alpnet.com, www.Berlitz.com, www.global sight.com, Lernout & Hauspie <www.lhsl.com>, www.Lionbridge. com, www.Logisoft.com, www.Logos-usa.com, www.translations.com, www. Uniscape.com

Machine translation providers: Alta Vista's Babelfish <www.babelfish.altavista. com>, www.IDC.com, www.e-Lingo.com, Lernout & Hauspie <www.lhsl.com>, www.Systransoft.com

Microsoft's global development page: www.eu.microsoft.com/globaldev/fareast/ fewinnt.asp

Microsoft Planning for and testing global software: http://www.microsoft. com/GLOBALDEV/Non%20mirror/back%20up/gbl-gen/INTREFNEW.HTM

Microsoft Windows Internationalization: http://www.microsoft.com/globaldev/ gbl-gen

Native-American-oriented Website: www.hanksville.org/NAresources/

Simplified English: userlab.com/SE.html

Internet users survey, Nua: www.nua.ie/surveys/how_many_online

Unicode: www.unicode.org/, IBM Unicode Glossary: www-4ibm.com/software,/de-veloper/library/glossaries/unicode.html

Women: wow.com, oxygen.com, chickclick.com

Resources

This section lists organizations providing information about international standards and intercultural issues.

• ACM SIGCHI Intercultural Issues Database
Gary Perlman maintains the ACM SIGCHI Intercultural Issues Database, which is summarized in (Perlman).

> Gary Perlman, OCLC Online Computer Library Center
> 5565 Frantz Road, Dublin, OH 43017
> Email: Gary Perlman <perlman@TURING.ACM.ORG>
> URL: http://www.acm.org/sigchi/intercultural/

• American National Standards Institute (ANSI)
This organization analyzes and publishes U.S. standards, including those for icons, color, terminology, user interfaces, and so on. Their contact data are the following:

> American National Standards Institute (ANSI)
> 11 West 42nd Street, 13th Floor, New York, NY10036
> Tel: 212-642-2000

Email: info@ansi.org
URL: www.ansi.org

- China National Standards Organization
This organization analyzes products to be imported into China. The contact data are the following:

China Commission for Conformity of Electrical Equipment
(CCEE) Secretariat
Postal address: 2 Shoudu Tiyuguan, NanLu, 100044, P.R. China
Office address: 1106, 11th floor, 2 Shoudu Tiyuguan, NanLu,
Beijing, P.R.China
Tel: +86-1-8320088, ext. 2659, Fax: +86-1-832-0825

- East-West Center
This organization, formerly funded by the U.S. Congress, is a center for technical and cultural interchange among Pacific Rim countries. The center's research and publications cover culture and communication. The contact data are the following:

East-West Center
1601 East-West Road, Honolulu, HI 96848-1601
el: 808-944-7111, Fax: 808-944-7376
Email: ewcinfo@ewc.hawaii.edu
URL: http://www.ewc.hawaii.edu

- Information Technology Standards Commission of Japan (ITSCJ)
Information Processing Society of Japan

Kikai Shinko Building, No. 3-5-8 Shiba-Koen, Minato-ku,
Tokyo 105, Japan
Tel: +81-3-3431-2808, Fax: +81-3-3431-6493

- Institute for International Research
This organization conducts research and organizes conferences about globalization and localization:

Institute for International Research, Inc.
2009 N. 14th Street, Suite 303, Arlington, VA 22201

Tel: +1-703-908-9010, Fax: +1-703-908-9018
URL: www.iir-ny.com

• International Standards Organization (ISO)
This organization analyzes and publishes world standards for all branches
of industry and trade, including standards for icons, color, terminology,
user interfaces, *etc.* Their contact data are the following:

International Standards Organization (ISO)
Geneva, Switzerland
Tel: +41-22-749-0111, Fax: +41-22-733-3430
Email: central@iso.ch
URL: http://www.iso.ch/

Of special interest is ISO 8601's international time and date standards.
Information about ISO's 8601 standards, and the particular one about time
and date standards, may be found at the following URLs:

http://www.aegis1.demon.co.uk/y2k/y2kiso.htm
http://www.roguewave.com/products/resources/exchange/iso8601.html

• Japan National Standards Organization
These organizations analyze and publish Japanese standards, including
those for icons, color, terminology, user interfaces, and so forth. Their contact data are the following:

Japanese Industrial Standards Committee (JISC)
Agency of Industrial Science and Technology
Ministry of International Trade and Industry
1-3-1, Kasumigaseki, Chiyoda-ku, Tokyo 100, Japan
Tel: +81-3-3501-9295/6, Fax: +81-3-3580-1418

• Localization Industry Standards Organization (LISA), The
This organization provides information and publications relevant to localization professionals:

The Localization Industry Standards Association
7, rue du Monastère, 1173 Féchy, Switzerland
Tel: +41-21-821-32-10, Fax: +41-21-821-32-19
URL: http://www.lisa.org

- Web of Culture, The
This Web site publishes news and resources related to cross-cultural communication:

    Ms. Eileen Sheridan, President
    113 Anita Road, Suite 6, Burlingame, CA 94010
    Tel/Fax: +1 408 273 6074
    URL: <http://www.webofculture.com>

- World Wide Web Consortium
This organization provides information relevant to globalization issues, including accessibility. Two URLs of interest are:

    http://www.w3.org/International, for information about internalization.
    http://www.w3.org/WAI, for information about accessibility.

Chapter 4

Synthesizing the Literature on Cultural Values

Emilie W. Gould
Rensselaer Polytechnic Institute, Troy NY

INTRODUCTION

I was having lunch with a Malaysian friend who was having problems with her doctoral advisor. "When I say I want to study culture and information systems, he says, 'That's too broad!' When I ask him what he means, he says, 'Culture is everywhere—national, ethnic, religious, corporate, family—but it's not consistent. You can't make useful generalizations based on culture. Even if you could, it would be impossible to develop separate interfaces for each different culture. Focus on the psychology of the individual user instead!'"

I said, "Yeah, I've had that problem too. A lot of people in the field of human–computer interaction (HCI) really resist the notion of culture because it seems like such a vague, anthropological concept. When HCI first evolved from experimental psychology, you could run a test in the usability lab and claim to know what every novice user might need. Then people designing cooperative work discovered qualitative research. But HCI is just coming to grips with culture. I tell people, 'I'm applying a set of well-accepted management theories to people's design preferences on the Internet.' They like that better."

Legitimating the study of culture has been a problem for anyone interested in the intersection of national difference, personal preference, and computer interfaces. We all know individuals from other cultures who seem more like us than like their compatriots. They make it hard to deny the claim that "People are where the interface is, not culture!" But you can say the same for any social science. People are different, yet general psychological profiles have been developed. Organizations vary, but in a regular way. Countries can be grouped together in useful typologies—democratic, socialist, authoritarian, and so forth. Within cultures, individuals cover a wide spectrum of belief and behavior but, in the aggregate, they cluster together and these clusters display a surprising amount of stability.

So, what is the best way to analyze national culture to improve human interface design? There are many ways of thinking about the problem, but generally some theories stress how cultures differ while others examine what they have in common—the "emic/etic" distinction of linguist Kenneth L. Pike (1954). Pike contrasted the universality of phon*etics* (language sounds) with the diversity of phon*emics* (interpretations of those sounds as coherent systems of communication). His analysis has been widely applied to social phenomena like writing and rhetoric that are based on a universal system like the alphabet but require selective interpretation by "insiders."

Cultural anthropology can provide useful insights into designing interfaces for specific countries, but theories from the field of intercultural communication are generally better for culturally diverse audiences. Most designers do not have the mandate to develop entirely different products for each national or ethnic market. Intercultural communication theory makes it possible for them to focus on a few crucial variations. Since the 1990s, when the concept of culture finally emerged as a concern for interface design, most design ideas have come from intercultural communication, most notably, in articles and workshops by Aaron Marcus (1993, 2001; Marcus & Gould, 2000) and in the essays in Elisa M. del Galdo and Jakob Nielsen's *International User Interfaces* (1996).

For the rest of this chapter, I (a) further discuss the advantages of intercultural communication theories for HCI design, (b) list key communication theorists whose work has heuristic value for the field, and (c) identify the most important variables for interface development.

THE ADVANTAGES OF INTERCULTURAL COMMUNICATION THEORIES FOR HCI DESIGN

Though cultural anthropology seems like it should be valuable for interface development, it has several limitations as a design guide. First, it tends to focus on specific cultures to the detriment of comparison and theory and, second, it often emphasizes authenticity over interaction. Some of these problems are inherent in its chosen methodology; others derive from its historical development as an academic field.

Many anthropological studies use "thick description" to record and highlight behaviors and attitudes that contrast with the anthropologist's own culture. Unfortunately, this approach emphasizes a degree of cultural uniqueness that may make it seem impossible to bridge cultural differences. Although each culture has its own symbols and patterns of action, not all differences are significant. Some of the interface development literature on cultural markers illustrates this problem. Tony Fernandes (1995) first (delightfully) introduced the problem of translating cultural symbols, taboos, and aesthetics into interface design but did not identify ways to determine which symbols were important. Wendy Barber and Albert Badre (1998) introduced the term *cultural marker* to define interface elements that "signify a cultural affiliation." They talked about developing a directory of flags, colors, and national symbols that would allow interfaces to be automatically recompiled with different cultural markers for different markets. Charles Sheppard and Jean Scholtz (1999) tried to test this idea, but their results were not significant. Their effort demonstrates the difficulty of working with an essentially atheoretical concept.

By contrast, the field of intercultural communication focuses on the regularities between cultures, identifying a few critical variables that help people craft appropriate messages and collaborate. In particular, the literature on value orientations reduces cultural differences to a manageable number.

In 1950, Florence Rockwood Kluckhohn first systematically compared cultures according to their values. The following year, her husband Clyde Kluckhohn (1951) used the term *value orientations* in an essay on theories of action. For the next decade, Harvard University sponsored Florence Kluckhohn and Fred L. Strodtbeck in the Values Project. This longitudinal study compared five distinct communities—Zuni, Navaho, Mexican, Mormon, and Texan—living near one another in the American Southwest.

Their results were published in 1961 as *Variations in Values Orientations*. The book's most notable contribution was its characterization of each community on the basis of five existential (and thus presumably universal) questions:

What is the purpose of human nature (and can people change)?	*Human-nature orientation*
How do people relate to nature (and the supernatural)?	*Man-nature orientation*
How do people manage time?	*Time orientation*
What is the purpose of human activity?	*Activity orientation*
How do people relate to each other?	*Relational orientation*

Three possible variations were assigned to each question. Taken together, these value orientations allowed Kluckhohn and Strodtbeck to identify opportunities for conflict and cooperation and to analyze each community equitably.

However, few other cultural anthropologists followed their lead. Through the rest of the decade, many studies continued to emphasize the distance between "traditional" societies and "evolved" cultures. Eventually, postcolonialist critics like Edward Said (1978/1994) began to document the role of academic disciplines in facilitating cultural imperialism. "Orientalism" and its ilk were seen to justify empire by emphasizing the inferiority of non-Western cultures. Anthropology came under attack for its focus on isolated, marginalized groups.

The field took these criticisms to heart; this crisis reformed its study of culture. Anthropologists today use a variety of techniques to avoid the unequal power relationships inherent in studies of "exotic" people. Members of the culture collaborate in the research (much as do union members in Participatory Design); lightly edited (but carefully selected) personal narratives and field notes may be published in place of abstract analysis; and, anthropologists increasingly investigate their own cultures.

But such work too often remains unitary and singular. In 1946, Ruth Benedict said that a culture might share 90% of its observances with its neighbors, but the differences, "however small in proportion to the whole, turn its future course of development in a unique direction" (1946/1989, p. 9.) Postmodernist and postcolonialist studies of "difference" often make it appear that nothing important is shared except the experience of oppression. Unfortunately, the moral imperative to avoid stereotyping and the desire to recognize "authenticity" are antithetical to a utilitarian focus on collaboration.

Intercultural communication came through a similar crisis but developed a more useful methodology. The field emerged after World War II and was shaped by the collapse of Europe, the emergence of the *pax Americana*, and political and economic globalization. In the 1940s, the difficulties facing many American diplomats and representatives of nongovernmental organizations led Congress to reorganize the Foreign Service and found the Foreign Service Institute (FSI). Linguists and anthropologists were hired to provide language and culture training. Historians and communications researchers collaborated with descriptive linguists like George L. Trager (who focused on nonverbal paralanguage) and Ray L. Birdwhistell (who developed kinesics). In 1959, the year after William J. Lederer and Eugene Burdick's *Ugly American* made the best-seller list, Edward T. Hall proposed a solution to the problem of "culture."

The Silent Language was based on examples from the FSI and showed how ignorance of cultural patterns and nonverbal communication underlay numerous cases of intercultural miscommunication. The book was easy to read and helped explain to millions of Americans why the country was being demonized while trying to do good. As a result of its popularity and the increasingly theoretical focus of his subsequent books, Edward T. Hall is often described as the founder of the field of intercultural communication. As he himself claimed, "The complete theory of culture as communication is new and has not been presented in one place before" (1959/1973, p. 32).

Intercultural communication was soon recognized as a separate communication specialty. Subsequent researchers looked for, and found, many patterns of cultural variation. Generally these variations were based on analyses of face-to-face communication between people from different cultures.

This interpersonal focus makes the field a good fit with HCI. According to Horton and Wohl's (1956) theory of *parasocial* communication and Reeves and Nass' (1996) "media equation," our interactions with "new media" are just an extension of our face-to-face interactions with one another. Consequently, theories that deal with communication between people from different cultures should transfer directly to communication between people from one culture using interfaces developed in another.

Hall has been followed by Edward C. Stewart and Milton J. Bennett (1972); John C. Condon and Fathi S. Yousef (1975); Geert Hofstede (1980; Hofstede & Bond, 1988); Harry C. Triandis (1975, 1983, 1995); William B. Gudykunst (1987; Gudykunst et al., 1996); Fons Trompenaars and Charles Hampden-Turner (1998); and David A. Victor (1992). Next, I review each of these researchers and discuss ways to apply their theories to interface design.

EDWARD T. HALL: CHRONEMICS, PROXEMICS, AND CONTEXT

Hall's background preparing diplomats at the Foreign Service Institute shaped his work. He felt that assignees should be tested for their psychological adaptability, trained in the appropriate language of their new country, and introduced to its formal culture. But, "training in language, history, government, and customs is only the first step. Of equal importance is an introduction to the nonverbal language of the country" (1959/1973, p. xiii). Until people understood this "language of behavior," they could not begin to be effective, and they needed to understand their own "silent language" to anticipate their reception.

In *The Silent Language*, Hall compared American cultural patterns with those of Japan, Germany, France, Greece, the island of Truk, and the Middle East. He focused on two nonverbal features that remain important in intercultural communication theory: time (chronemics) and space (proxemics).

In *Beyond Culture* (1976), he identified another factor, which governs people's responses to verbal discourse, that is, context.

Hall's discussion of chronemics is quite detailed. Time is organized in two ways. The first is the subjective division of time into technical, formal, and informal systems. The second is the connection of time with activity.

Technical time reflects the underlying physical context of our experience of time. The solstices, phases of the moon, sunrise, and atomic decay can all be used to separate time into units. Formal time is often based on technical time. It consists of conventional systems of measurement that help people plan, schedule, and manage time. Formal time also leads to a sense of determinacy—the feeling that events are linked and causal. However, nothing requires a day to have 24 hours or 7 days to make a week.

Informal time is cultural and situational. It exists side by side with technical and formal time and frequently uses the same vocabulary, but it involves culturally different perceptions of the rate time passes. For instance, if I say, "I will see you in 5 minutes" in Germany, I need to hurry to meet you. If I say the same thing in Ecuador, you will not be concerned if I don't show up for an hour. Informal time also deals with the significance of events in the past, present, and future. Hall noted that people in the Middle East and South Asia are more likely to ground decisions in historic events than are people in the United States.

The connection between time and activity led Hall to his notions of *monochronic* and *polychronic time*. In monochronic societies like the United States, Canada, and northern Europe, people do one thing at a time. They prefer to work sequentially and try to resist interruptions. In polychronic societies like Latin America, the Middle East, and parts of Asia, people work on several things at once. This may mean promising to meet different people in different places at the same time or working on two or three projects simultaneously. Monochronic time enhances determinacy but, in polychronic time, things tend to "happen."

Hall's second contribution to intercultural theory is *proxemics*, the social use of space. Once again, he analyzed the expression of this construct in different cultures in terms of technical, formal, and informal systems. Physical constraints (mass, edge, volume) underlie technical systems of space; the process of architecture gives rise to various formal systems; and people interpret informal patterns of proximity and arrangement to determine status and group orientation.

With *Beyond Culture* (1976), Hall developed the idea of *context*. Although this construct is also used in literary theory for situating text and verbal discourse, Hall wanted to highlight the importance of nonverbal communication. In high-context societies, features of the social system shape meaning through chronemics, proxemics, and other nonverbal behaviors; a message's actual text is secondary. In low-context societies, the information is explicitly expressed in the text of the message.

The United States tends to be low-context; Americans generally say exactly what they want you to understand. Many Asian societies (Japan is the classic example) are high-context; the message is inherent in the occasion, the physical setting, and the relationships between the participants. People within the culture know how to interpret each other's intent, but many American business people have learned the hard way that "yes" (said politely but unenthusiastically) really means "no."

Hall as a Guide for Interface Development

How do Hall's constructs relate to interface design? First, his definition of technical, formal, and informal systems can be mapped to cultural perceptions of applications and the Web. Computers also have certain technical limits based on hardware and software, but the formal layout of screens (and sequences of screens) can still be read in different ways. For instance, the "architecture" of a web page may differ in terms of color, mass, and balance between verbal and visual elements. Many American web sites currently sport subdued hues, arrange content in explicit hierarchies of indentation, and emphasize text over graphics. Web sites from other countries appear visually striking if they upset American expectations by using vivid colors, an asymmetric arrangement, or more graphics than text. Appropriate content may also vary. Text that lacks formal historical grounding will be unpersuasive to people who rely on tradition to justify their actions. Moreover, presentation and content may interact. Pictures that are read individually in one country can become part of an unintentional collage of meaning in another.

Second, Hall's notion of time has implications for interaction. Different applications may express monochronic or polychronic time. A monochronic application can support a user's time sensitivity by focusing narrowly on a task or (somewhat paradoxically) by providing a wide range of choices to reduce the search hierarchy. A polychronic application can present less structured options or support incomplete sequences of tasks. For instance, a user may look up a train schedule on the Internet but expect to return to the home page (or go to the station) to complete the transaction.

Third, Hall's notion of context is important because it says that high-context users may not focus on the explicit verbal message; thus, a developer cannot depend on written text for clarity. Context may be expressed in the nonverbal elements of the screen, or context may be out of the developer's control and inherent in the status of the software provider or in local procedures that constrain the user's access to the application. As a result, a web site for a low-context audience may not be appropriate for a high-context one. A site may lack credibility because it does not provide branding or information on the social structure of the sponsoring organization. A site can also fail because it does not recognize the social status of the user or because it addresses a high-status user with insufficient respect. Under certain circumstances, applications can actually be "too easy." Lee (2000) found that Korean subordinates prefer to fax messages to their superiors because e-mail did not require enough effort to show sufficient respect.

EDWARD C. STEWART AND MILTON J. BENNETT: SUBJECTIVE AND OBJECTIVE CULTURE

Edward C. Stewart and Milton J. Bennett believed that communication style and cultural values arise from the process of human perception. They wrote their classic book on *American Cultural Patterns* (1972/1979) to help U.S. citizens working abroad. Before such sojourners can properly interpret messages and events in other cultures, they need to recognize how they have been conditioned by their own. Like Hall, Stewart and Bennett believed that intercultural communicators must be culturally self-aware to be effective.

Culture has two fundamental aspects. Objective culture is social and material, encompassing a society's political and economic system, aesthetics, customs, art, architecture, and institutions. Subjective culture is psychological, reflecting on society's values, expectations, theories of action, and patterns of thought. Objective culture is tangible (or at least easily talked about); subjective culture tends to be experienced unconsciously.

Cultural clashes arise from mismatches in the process of human perception. Subjective culture can be brought forward into mindfulness by understanding how the "deep mind" interprets its environment. An understanding of the different subjective levels of thought helps identify the sources of cultural miscommunication.

All people have the same senses, but people from different places often draw different conclusions from the same stimuli. Their minds are conditioned to filter out or ascribe different messages to similar sensory information. For instance, people may convert the same stimuli into different percepts. A small mammal that barks is food for one person, a pet for another. As part of the process of distinguishing figure and ground, different people foreground different information for their idea of a "dog."

At the next level of mindfulness, people develop patterns of thinking. Personal development within a family, community, and culture helps shape their cognitions. Memory, emotion, stories people have read, and stories they have heard from others bring to mind dinners in Shanghai or walks in the park. This is the stage at which "the mind extends its mastery of objective reality by creating a cognitive, or subjective, reality where meaning is assigned and relationships elaborated" (Stewart & Bennett, 1972, p. 22). Patterns of thinking apply to all types of sensory data—graphic and nonverbal, as well as text and language. Screen layout and expectations of interaction are as subject to the values and assumptions of culture as are words and pictures.

Complex symbol systems make up the deepest layer of the Stewart and Bennett model. These systems—pictorial, musical, linguistic, and mathematical—give rise to metaphor, grammar, and semantics. People learn these systems within the larger societies that use them. (They can also be learned formally but it takes an insider to explain that = "dog.")

So, how does language influence values? The two are intertwined.

Stewart and Bennett acknowledge a debt to Benjamin Whorf and the structural linguists. In analyzing patterns of language use in the United States, they tie grammar to behavior:

- The subject/predicate form of English is linked to a heightened sense of causality ("*it* as implied agent").
- The use of dichotomies and implied negative constructions is linked to critical thinking.
- Modifier/noun patterns are linked to a preference for process and precision.

These syntactical elements enhance active, independent behaviors and work against indirection or collaboration. People who use such language are likely to be part of or develop an individualistic culture.

Stewart and Bennett make a good case for looking beyond the literal content of visual imagery and language and grounding cultural analysis in perception and the underlying grammar of symbol systems. More than 30 years later, Richard E. Nisbett (2003) has identified similar cognitive differences that affect the way Westerners and Asians "read" pictures—and computer screens.

Stewart and Bennett as a Guide for Interface Development

Most computer companies base their internationalization efforts on the premise that all users perceive computer systems in the same way. If everyone engages the system at an equivalent semantic level, then language localization should be sufficient to address national differences. At worst, a company might be required to change certain metaphors, for example, a U.S. mailbox becomes a British post box.

But Stewart and Bennett's distinction between objective and subjective culture increases the possibility that companies must go beyond surface text and images. The grammar of the interface is hard to recognize but may be fundamental in cultivating trust and acceptance. Design elements like alignment, proximity, repetition, and contrast (Williams, 1994) affect the massing of text on a page and the overall balance of the screen. Such layout features should not be ignored when localizing screens because they can contribute to intercultural miscommunication. For instance, text that is read from left to right trains users to look for important information in the top left and lower right quadrants of a screen. But bidirectional text is read from right to left and Asian scripts may be read from top to bottom. So where should the most important information be placed? When translating a process, the order of the graphics should be reversed.

HCI researchers need to identify similar cultural differences in perception and interpretation. A grammar of visual rhetoric, like that of Gunther Kress and Theo van Leeuwen (1996), will not be stable across cultures. An image of a young man gazing back from the screen may promote engagement with the interface in Britain but intimidate a new user in East Asia. Cultural messages are embedded throughout the user interface.

Charles Kostelnick (1995) has also emphasized the link between perception and aesthetics. Interface designers tend to be so embedded in the visual language of modernism that they are incapable of recognizing their own bias. It is necessary to understand how people in different cultures experience an interface to address their hidden (and seemingly irrational) motivations. Such cultural knowledge must include their artistic preferences as well.

Nancy Hoft (1996) championed the value of Stewart and Bennett's "metamodel" for constructing narrowly targeted cultural models for interface localization. Designers should first build a model of their own culture to have a standard for comparison. Then, thinking about unconscious levels of perception will help determine what parts of the interface to focus on. Fundamental differences in objective and subjective culture will mandate extensive change in localized screens and content.

By using a model of culture, Hoft promised that designers could identify:

- Global information that can be put into the interface without requiring future translation.
- Cultural bias in the existing application.
- Parts of the application that should be localized for a specific culture.
- Compelling cultural metaphors and cultural markers.
- Potential cultural problems.
  (In addition, cultural models provide guidance for developing appropriate methods for usability evaluation.)

However, Hoft might have taken Stewart and Bennett's perspective even further. She talks about unspoken and unconscious rules in her own "Iceberg" metamodel and highlights Hall's contributions to decoding nonverbal communication, but she does not fully apply these insights herself. For instance, she recommends identifying international variables for cultural models by surveys and observation, but the first depends on self-report and the second on an observer being able to recognize the triggers for unconscious behavior.

JOHN C. CONDON AND FATHI S. YOUSEF: EXTENDING THE VALUE ORIENTATIONS APPROACH

In 1975, John Condon and Fathi Yousef published *An Introduction to Intercultural Communication*. As their editor noted, this was the first survey to pull together the diverse themes associated with interpersonal, nonverbal, and speech communication; combine them with value orientations; and provide the reader with a single, unified explanation of cultural differences in social structure, language, and argumentation. (Even Hall developed his theories piecemeal.)

Condon and Yousef cited everyone from Hall, Bateson, and Goffman (for interpersonal communication) to Birdwhistell (nonverbal communication and paralanguage) to Toulmin and Aristotle (rhetoric), Sapir, Whorf, and Chomsky (linguistics), and McLuhan (mass communication). But the focus of their book is their extension of Kluckhohn and Strodtbeck's five existential questions. They defined 25 value orientations under the headings of:

- *Self*—individualism/interdependence, age, sex, and activity.
- *Family*—relational orientations, authority, role behavior, and mobility.
- *Society*—social reciprocity, group membership, intermediaries, formality, and property.
- *Human Nature*—rationality, good and evil, happiness, and mutability.
- *Nature*—relationship of humans to nature, ways of knowing nature, structure of nature, and concept of time.
- *The Supernatural*—relationship of humans to the supernatural, meaning of life, providence, and knowledge of the cosmic order.

These categories are best represented as a Venn diagram (see Fig. 4.1) with overlapping spheres representing self, society, and nature with the intersections representing family, human nature, and the supernatural.

Condon and Yousef's (1975) definition of cultural values is flexible. The number of value orientations is not absolute. There could be more—or fewer:

> Each of the value orientations ... arises from empirical data obtained from many cultures. The greater danger is not that of hypothesizing irrelevant categories but in not including enough. Since the scheme is not intended to be exhaustive, this is not a major problem; one is always free to add more or, if desired, to combine categories or to ignore some entirely. (p. 58)

Cultural values develop early in life. Half of the values deal with "what is thought to be good" and the other half with "what is thought to be true."

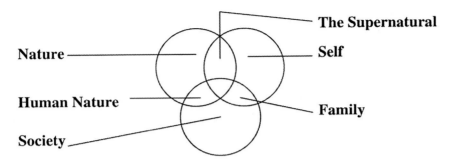

FIG. 4.1.   Condon and Yousef's model of culture.

Though standards of goodness are often easier to recognize and articulate, diverging beliefs about truth tend to have more profound consequences for intercultural communication. Opposing notions of reason, pleasure, change, and purpose may hinder individual perception and prevent people from agreeing on common logics, aesthetics, strategies, or goals.

Condon and Yousef as a Guide for Interface Development

The expansion of Kluckhohn and Strodtbeck's value orientations to social structure, language, paralanguage, and argumentation moves their theory from existentialism into the realm of pragmatic communication. Because three options are associated with each value, a broad range of behaviors is described by the 75 heuristics.

Values like interdependence and the use of specialist intermediaries could have a direct bearing on deciding whether to make agents a prominent part of an interface. Preference for youth versus age, women over men, or individuals with tighter or looser relationships to their families might influence representations of users. The activity orientations and relationship of humans to nature (human domination, harmony, or the supremacy of nature) should be factored into any decision to develop utilities to change application defaults and system parameters.

But the greater importance of Condon and Yousef's conceptual scheme lies in their emphasis on rhetoric and persuasion. How narrowly should an application focus on specific tasks? Should an interface be organized deductively or inductively to appeal to people from a different culture? What kind of information will be persuasive? Once developers recognize that their own patterns of rationality are not universal, there is the possibility of designing new interfaces with more appeal to those voluntarily on the opposite side of the digital divide.

Condon and Yousef also highlight the importance of communication roles. People communicate with images of each other: You communicate with an image of me and I communicate with an image of you. In addition, my image of "me" also influences my expectations for interaction with your image of "you." Walter Ong (1975) once said that an author's audience is always a fiction. Interface designers are no less guilty of inventing their users.

In fact, our projections are often hyperpersonal. Joe Walther (1996) found that unacquainted people communicating at a distance make more extreme, and generally more positive, judgments of one another than people communicating face-to-face. We contribute to the digital divide when we have unrealistic and uncritical expectations that everyone who uses a computer must be much like us. Unfortunately, when they are not, their reactions can become overly negative. In general, bad experiences diminish willingness to try new software. Such problems take on a geopolitical dimension because computer interfaces and Internet web sites function as communication partners in themselves. When television was new, Horton and Wohl (1956) described people's strong engagement with TV stars and other media figures. Reeves and Nass (1996) found no difference in classic social psychology experiments rerun with humans and a computer surrogate. People who interact with our software feel that they are interacting with our culture. Poor usability may be seen as confirmation of a vendor's (and a culture's) lack of respect.

Thus, our interpretation of users from other countries is often based on unrealistic images of them. And their (often negative) projections of us, based on their interactions with our products, may be equally unfounded. With the best of intentions, people in HCI tend to underestimate the effects of culture even as we struggle to close the digital divide.

GEERT HOFSTEDE: THE DURABILITY OF THE "SOFTWARE OF THE MIND"

Hofstede's (1980) theory of universal cultural dimensions has become the best known, and probably most applied, intercultural communication theory. Various researchers have applied it to leadership, motivation, compensation, and teamwork within domestic and multinational firms, and Nancy Hoft (1996) used it in her article on developing a cultural model.

Hofstede based his work on a series of surveys from 1968 to 1972 with IBM employees from 72 national subsidiaries; after factor analysis, four dimensions explained half the differences between the 116,000 respondents. Not only were the factors supposed to be universal, but Hofstede claimed that they were highly resistant to change because people learned them as children. Adults might be exposed to opposing values and superficially adapt, but they could never erase this "software of the mind."

Aaron Marcus (chap. 3, this volume) thoroughly describes these dimensions:

- *Power Distance*: the extent to which less powerful people expect and accept that power is distributed unequally.
- *Individualism vs. Collectivism*: the extent to which people are integrated into tight social networks.
- *Masculinity vs. Femininity*: the relative desirability of material success versus quality of life and of assertive versus modest behavior.
- *Uncertainty Avoidance*: the extent to which people tolerate ambiguity and risk or feel threatened by change.

In addition, Marcus demonstrates how they can be mapped to a model of the user interface to generate culturally appropriate designs. (See also Marcus & Gould, 2000.)

Hofstede's dimensions have become a touchstone for comparisons of national differences because of the size and scope of the original survey, but they have also been attacked. Three major criticisms have been raised against Hofstede's four factors:

1. IBM employees are not representative of any national culture. As employees of a multinational corporation with a conservative (in those days) corporate culture, they should not have been used to sample national traits.
2. The factors are not stable. Cultures change over time and even IBM is not the company it was in 1972 when Hofstede finished his research.
3. Hofstede's survey was itself biased. Because he developed and tested his questions in Europe and the United States, and administered them in English, they were permeated with Western values.

Hofstede acknowledged that IBMers are not representative but rejected the implication that they can not be used to identify cultural values. Instead, he claimed that their very bias makes them a functionally equivalent sample. That is, IBM's distinct workplace culture gives employees so much in common that any differences between them must be the result of differences in national culture.

Hofstede also rejected the criticism that his factors might change over time. Instead, he emphasized the stability of childhood acculturation. Recent work by Hofstede fellow and colleague, Marieke de Mooij (2001), reiterates this point. But Hofstede did admit that his country rankings omit some nations (e.g., People's Republic of China) that have become more important in the three decades since he did his research.

Hofstede also accepted that his original survey on employee attitudes does have a Western bias. Michael Bond (a Canadian-born professor in Hong Kong) convinced him to administer the Chinese Value Survey, which had been developed by Taiwanese and Hong Kong researchers in the early 1980s. After reviewing the factor analysis, Hofstede and Bond (1988) found a fifth cultural dimension: long- versus short-term orientation (also known as Confucian dynamism, the acceptance of traditional Confucian values).

The Chinese Value Survey supported the validity (and universality) of power distance and individualism and collectivism. However, it modified the definition of masculinity to include more "chivalric" values like patience, courtesy, kindliness, and compassion.

This redefinition highlights a major problem with the third of Hofstede's dimensions. Notions of masculine and feminine behavior vary widely and are themselves culturally determined. A more useful way to interpret this dimension might be to examine the extent that people are bound by general social roles. In masculine societies, it is difficult for men to look after children, but it is also difficult for anyone to change their religious or political affiliation. In feminine societies, people are less bound to social roles in general.

More recent research repeats these criticisms of Hofstede's methodology while discovering surprising confirmation for his strongest dimensions. For instance, Shalom Schwartz (1994) published an alternative set of value dimensions developed from a new sample of 41 cultural groups in 38 nations.

As shown in Fig. 4.2, his factor analysis juxtaposes two qualities found in individualistic societies—Intellectual Autonomy (self-direction) and Affective Autonomy (stimulation and hedonism)—with the Conservatism (concern for security, conformity, and tradition) found in homogeneous (collectivist) societies. Further, Schwartz was able to identify factors associated with Self-Enhancement that contrasted with values promoting Self-Transcendence.

The two Self-Enhancement factors are Hierarchy (the legitimacy of power hierarchies) and Mastery (self-assertion, self-promotion, and active control of the social environment)—factors similar to Hofstede's Power Distance and Masculinity. The two Self-Transcendence factors are Egalitarian Commitment (the motivation for altruism and volunteer service in individualistic societies) and Harmony (with nature)—factors that expand Hofstede's notion of Femininity.

Schwartz better explains why individualists often act collectively, for instance, how Sierra Club members in Los Angeles and subsistence farmers in traditional African villages can share a common environmental perspective. But, overall, Hofstede's dimensions hold up surprisingly well.

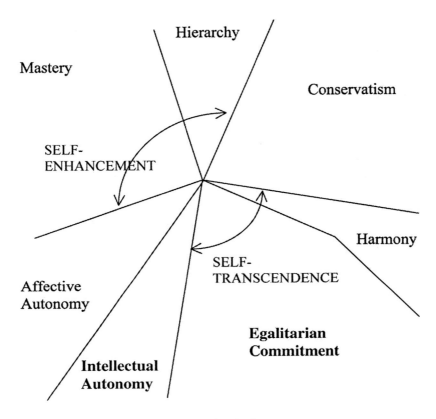

FIG. 4.2.   Shalom Schwartz's culture-level value structure.

Hofstede as a Guide for Interface Development

Hofstede's tables of cultural differences provide a methodology for antici-
pating cultural differences. Each of the 50 countries (and three regions) in
his original survey is ranked in terms of the original four work orientations,
with another 20 listed for Confucian Dynamism. Thus, if you are a Canadian developing a web site for South Koreans, you can determine that your
audience will have higher power distance, lower individualism, greater
masculinity, higher uncertainty avoidance, and a more long-term orientation than a domestic audience.

   When you associate interface elements with these dimensions, you may see
some overlap. Support for mastery (masculinity) has much in common with
allowing people to learn sites by trial and error (low uncertainty avoidance).

Hofstede admits that most of his factors are correlated with one or two others, but considers this redundancy helpful, not problematic. Not all cultural differences are equally critical for each country. Only significantly different cultural orientations warrant attention, though each can percolate throughout the interface.

For example, Canada and South Korea are quite different, and a Canadian designer might need to develop a radically different site for a Korean audience. But a German designer modifying a domestic web site for Dutch customers would discover that the two countries share similar values for power distance, individualism, and uncertainty avoidance; only masculinity (Germany) and femininity (the Netherlands) are ranked far apart. As a result, the German could focus on the representation of gender and social roles and leave the rest of the site alone.

HARRY C. TRIANDIS: ADDING CULTURE TO SOCIAL PSYCHOLOGY

In the same decade that Hofstede started researching culture, Harry C. Triandis began to consider its effects on social psychology. He first examined the interplay between norms, attitudes, and behaviors. By the end of the 1960s and early 1970s, he was looking for differences in group identity to explain people's actions—analyzing subjective (as opposed to material) culture in Greece, race in the United States, and social distance between American, German, and Japanese students. He linked the influence of "in-groups" and "out-groups" to the issue of culture; in fact, he saw culture as the ultimate group.

At the time, social psychology was coming to realize that many of its topics were culture-bound and had little explanatory value. Studies replicated with people from different ethnic and national backgrounds showed radically different results. Variables often measured trivial amounts of difference.

In 1975, Triandis wrote an influential article that reviewed these problems and urged the field to focus on higher order constructs, like association and dissociation, whose expression could be compared across cultures. He acknowledged that intercultural studies would be difficult to develop because they require broad-based collaboration, multiple methods, and customized test materials for each culture, but he encouraged people to work cross-culturally.

Shortly thereafter, Triandis promoted models of universal social behaviors: "Eventually social psychological theories will be computer programs" (1978, p. 14). In another article, "Dimensions of Cultural Variation as Parameters of Organizational Theories," he urged social scientists to apply cultural values (most from intercultural communication) to then-current theories of productivity, job satisfaction, and leadership:

Organizational theorists naturally develop theories about the part of the social world that they know and understand. As a result, many such theories are
inappropriate and inadequate when applied to other cultures. We create organizations that work well for us, and for people like us. But these organizations are usually a failure for people with a different perspective.... What is
likely to be most effective in one culture is often different from what is most
effective in another. (1983, pp. 140–141, 165).

Triandis' promotion of these research goals makes him foundational to
the field of cross-cultural psychology. In 1976, he was elected president of
the International Association of Cross-Cultural Psychology (itself founded
only 4 years earlier.) In the 1980s, he edited the multivolume *Handbook of
Cross-Cultural Psychology*.

Over the last 15 years, he has taken his own advice and focused on a single cultural syndrome: individualism and collectivism. (This was one of
Kluckhohn and Strodtbeck's original value orientations and is the most
common cultural dimension in intercultural communication theory.)
Drawing on philosophy, political science, religion, and history, as well as
communication, anthropology, and psychology, Triandis has examined individualism and collectivism in depth. Much of his work considers such puz-
zling aspects as the strength of in-groups, the paradox of individualists in
collectivist societies (and vice versa), and interactions with power distance.

In reconceptualizing the dimension, Triandis (1988) asked himself why
people in collectivist societies are disproportionately influenced by their
in-groups. He decided that people in individualistic societies have greater
freedom of action and the opportunity to belong to more groups than people in collectivist societies. As a result, no single group dominates them,
though they often spend more time overall conforming to group norms. By
contrast, people in collectivist societies have fewer choices and belong to
fewer groups. Because their societies are also more homogeneous, these
groups (family, religion, and state) tend to hold similar values and reinforce
one another. Consequently, groups exert more control over people's actions. People in collectivist societies learn to be cautious and gather a great
deal of information about new individuals and situations because group relationships make such large demands on their time, energy, and autonomy.
However, outside their groups, people can act quite independently.

A year later, noting that people function on an individual (not cultural)
level, Triandis (1989) incorporated a theory of social identity into the dimension. People growing up in different cultures have more or less
well-developed private, public, and collective selves that they rely on for
appropriate social responses. In different situations, one self usually dominates and influences the development of personal relationships and patterns of communication. The individual level of individualism is *idiocentric*

and the individual level of collectivism, *allocentric*. These individual-level values may contrast with the overall culture-level values assigned to a specific country.

Triandis (1995) even found a way to redefine power distance as a subordinate dimension of individualism and collectivism (building on Hofstede's recognition that the two strongly covary.) *Horizontal* (low power distance) *individualism* is linked to the freedom to define and pursue one's goals, and *vertical* (high power distance) *individualism* to competition and accomplishment. *Horizontal collectivism* is tied to interdependence, in-groups, sharing, and consensus, whereas *vertical collectivism* is associated with service and the internalization of group norms.

Each category has its negative side. With horizontal individualism can come loneliness; with vertical individualism, personal stress. Horizontal collectivism can lead to distrust of people outside the group; vertical collectivism, to the emergence of excessive bureaucracy and rigid observance of rules. People are most satisfied with their lives when there is a good fit between their personal self-concepts and their social environment.

Triandis as a Guide for Interface Development

Triandis' skepticism about the ability of Western social science to identify universal principles of human behavior or universal methods of human research is a critique HCI should consider. All science is embedded in culture—computer science and interface design no less than social psychology. Triandis urged his own field to embrace collaboration and multi-site testing as the solution. This methodology is one that HCI should also adopt.

In addition, Triandis strengthened the explanatory value of individualism and collectivism. His explanation of the power of in-groups suggests that interface designers must work harder to establish trust with new users from collectivist cultures. His division of individualism and collectivism into multiple levels (individual, organizational, and societal) explains how a person born in Japan can be more independent than one born in the United States. (A society may be predominantly individualistic or collectivist, but individuals within that society can vary.) His theory of social identity acknowledges that a person (or culture) can be simultaneously individualistic and collectivist.

These refinements make individualism and collectivism elements in a personal repertoire and improves our understanding of them as design heuristics. Moreover, it improves them as experimental variables. Culture-level dimensions are particularly problematic in usability testing with the small number of subjects in the average usability lab. Many studies based on Hofstede's constructs have returned ambiguous results because the researchers assumed that everyone in a culture shares the same cultural orientation. Individual-level variables provide better information about their users and more realistically reflect the range of beliefs within cultures.

Hazel Rose Markus and Shinobu Kitayama (1991) incorporated Triandis' notions into their theory of independent (individualistic) and interdependent (collectivist) self-construals. These personal attitudes are easily measured with Singelis' (1994) Self-Construal Scale (since modified) by Cross, Bacon, & Morris, 2000) and make a better research variable for HCI than assignments based on Hofstede's country rankings.

WILLIAM B. GUDYKUNST: BUILDING A UNIFIED THEORY OF COMMUNICATION

William B. Gudykunst's work in the field of communication parallels Harry Triandis' advocacy of a cross-cultural perspective for social psychology. Like Triandis, Gudykunst first saw cultural differences as fundamental but now works to reconcile them within higher level constructs. He too has explored the importance of social identity. However, unlike Triandis, much of his work is based on the analysis of specific communication contexts. Instead of focusing on a single cultural syndrome, Gudykunst has investigated the effect of culture on initial encounters and relationship building. Recently, he has also looked at communication style.

Gudykunst's practical experience as an Intercultural Relations Specialist for the U.S. Navy in the early 1970s first helped shape his career. "While conducting intercultural training in Japan, I thought intercultural communication (i.e., communication between people from different cultures) was different from intracultural communication (i.e., communication with members of our own culture)" (1994, p. ix).

Gudykunst's first papers show this training bias. He examined the way people from different cultures encounter and learn to trust one another by applying Hall's and Hofstede's dimensions to classic communication processes like homophily (similarity) and uncertainty reduction (1983a, 1983b). But by 1984, he and Young Yun Kim concluded that intercultural and intracultural communication vary along a single continuum from total strangeness to total familiarity.

Initial interactions are particularly critical when people from different cultures meet because they are likely to have less in common. They begin by disclosing information about themselves and asking questions. These strategies help develop attraction by identifying common values, discovering common relationships, reducing uncertainty, and increasing attributional confidence. The process of interaction becomes an opportunity for coordination and affiliation.

At first, social stereotypes and information about group memberships are particularly important. As the relationship develops, people make more predictions based on knowledge of each other's personal characteristics. Eventually, the need for similarity decreases (old friends tolerate greater differences). This pattern is consistent across cultures, but the information requirements differ. People from high-context cultures ask more questions about family and background; people from low-context cultures are more immediately sociable, expressive, and direct.

Gudykunst and Tsukasa Nishida (1986) found a correlation between Hofstede's dimensions and levels of intimacy in six types of relationship. Gudykunst, Elizabeth Chua, and Alisa J. Gray (1987) confirmed that people look for greater evidence of similarity when beginning new intercultural friendships. In 1988, Gudykunst pulled together his work on individual and group representations and relationship development and published it as Anxiety/Uncertainty Management (AUM) theory (updated in 1993 and 1995.)

In addition to collapsing inter- and intracultural communication, Gudykunst can also be credited with introducing new theories from Europe into interpersonal communication. In 1986, he edited *Intergroup Communication* and wrote an introduction with Tae-Seop Lim:

> This book is intended to bring together the study of intergroup relations in Europe and the study of interpersonal communication in the United States. Both areas have developed separately, with little borrowing of concepts or theories. There are advantages to both areas, however, in their integration. It is through communication that intergroup relations are established and communication plays a central role in their improvement or deterioration. The intergroup relations perspective brings a focus upon group membership and its influence on the study of interpersonal communication that virtually has been ignored to date in the United States. (1986, p. 1)

Individuals come together in groups for a variety of reasons (perceptions of similarity, common interests, interdependence, or the desire for interaction.) Members learn to identify with one another and this social identity becomes part of their self-concept. People can belong to many groups and develop multiple social identities. Personal identity makes up the remainder of their self-concept.

When social identity is dominant, intergroup communication takes place. When personal identity is dominant, interpersonal communication takes place. Situations like labor negotiations activate both types of identity; anonymous encounters on the street activate neither. Gudykunst and Tae-Seop Lim's (1986) model (see Fig. 4.3) highlights the influence of communication context on identity. The contact situation often determines whether social or personal identity is made salient.

The final theme in Gudykunst's work has been his research on communication style. He used Hall's theory of high- and low-context communication

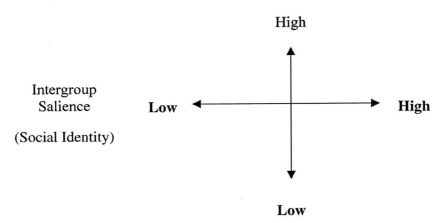

FIG. 4.3.   Interpersonal and intergroup communication.

in his work on initial interactions and relationship development, but he has also looked for the predictors of style. In 1988, Gudykunst, Stella Ting-Toomey, and Chua developed a model of interpersonal communication. This model shows how culture indirectly mediates communication style and content through social cognitive processes, situational factors, affect, and personal habits.

A few years later (1996), the model was simplified by incorporating Markus and Kitayama's self construals (see Fig. 4.4).

Gudykunst and Ting-Toomey collaborated with Yuko Matsumoto, Nishida, Kwangsi Kim, and Sam Heyman to survey high- and low-context styles in the United States, Japan, Korea, and Australia. When defined at the level of culture, individualism and collectivism only partially predicted communication style; however, when defined on an individual level (independent vs. interdependent self construals plus personal values), the dimension fully predicted communication style. People who considered themselves unique and independent used a low-context style, whereas people who considered themselves part of larger social relationships used a high-context style.

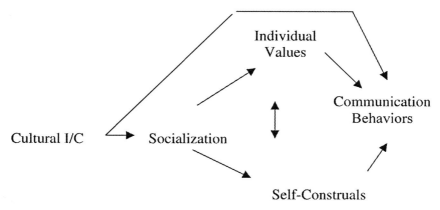

FIG. 4.4. Model of communication styles.

Gudykunst as a Guide for Interface Development

Gudykunst's research on initial encounters, identity, and communication style has direct relevance for HCI. In addition, it suggests some interesting intersections between intercultural and computer-mediated communication.

"Meeting" new software (or a new web site) invokes the same process of prediction, attribution, and uncertainty reduction that people use when meeting a stranger. Therefore, interface designers need to provide appropriate relationship-building information because people will make these attributions even when the uncertainty reduction process is unacknowledged. Group memberships are particularly important information for people who use a high-context style; personal expressiveness is particularly persuasive for people who prefer low-context communication.

Global access makes it likely that Web designers must include both types of information. The Web also provides some interesting technical opportunities for enhancing subsequent interactions. Cookies permit personalization of a site visited before. Usability, task orientation, and efficiency can be optimized to show consideration to low-context communicators. Interactivity by itself will enhance liking, that is, as long as the experience is positive. (See Rafaeli, 1988; Rafaeli & Sudweeks, 1998.) But new users may take different amounts of time to develop a sense of trust. People from high-context cultures are likely to need more information from the interface before they commit themselves; on the plus side, they are probably willing to explore more of the web site or application to find it.

Gudykunst and Lim's discussion of the salience of social and personal identity also has some interesting implications for HCI. Different contexts of use, for example, learning online presentation software at work, buying printer paper from the cheapest supplier on the web, or picking a university from a distance, may make different parts of one's identity salient. People in a work environment make the effort to learn a new application because it is a necessary part of their corporate identity. People engaged in an online transaction that involves little risk and no status or sense of participation simply send their order to the lowest vendor. But people researching an organization that requires a long-term commitment (or carries high risk) may need different information, depending on whether their preference is for high- or low-context communication. For instance, an Asian technician deciding to come to Europe for a training program could require more text on the history and management of the institute than a North American. But the same Asian would not need that level of information when picking a skiing vacation. Considering the effects of an application on social versus personal identity should become a new design heuristic.

Several computer-mediated communication theories focus on the same issues. Initial interactions and social identity come together in Martin Lea and Russell Spears' (1991) SIDE model (Social Identity model of Deindividuation Effects). This model suggests that people working together across computer networks make attributions about one another despite the lack of nonverbal cues present in face-to-face meetings. When their mutual group identity is enhanced and made salient, they will collaborate productively; when their individual identities are enhanced, they may "act out" (deindividuate) and ignore the needs of others. Joseph Walther and Judy Burgoon (1992) found that people in computer-mediated teams achieve the same level of liking and support as people in face-to-face groups, but it takes more time. Due to Walther's hyperpersonal effect, people may even like one another more because their impressions are based on positive projections and stereotypes. In deciding what kind of information to front-load on a home page, web designers need to consider whether people will be visiting the site once or returning repeatedly and whether they need to gather information to personalize subsequent interactions.

Finally, Gudykunst's work on communication style with Matsumoto, Ting-Toomey, Nishida, Kim, and Heyman (1996) has direct application to interface development. Eight factors emerged when respondents from Australia, Japan, Korea, and the United States were asked to rate 158 communication behaviors related to high- and low-context communication:

1. Certainty in inferring others' meanings, intentions, needs, and feelings (associated with individualism because it reflects a conscious strategy).

2. Use of indirect or ambiguous communication (collectivist).
3. Interpersonal sensitivity, tact, and preference for harmony (collectivist).
4. Use of dramatic communication (individualist).
5. Openly expressed use of feelings to guide behavior (individualist).
6. Personal openness and willingness to initiate communication with others (individualist).
7. Precision in communication (individualist).
8. Positive perceptions of silence in communication (collectivist).

Some of these styles reflect the way the respondents process information; others, the way they expect information to be delivered. The eight factors were correlated with independent and interdependent self construals and with personal values derived from individualistic and collectivist cultures. Because people can have both independent and interdependent self construals, the authors note that specific situations may activate opposite self-concepts. For instance, people might use a different style communicating with an in-group than with an out-group they had no interest in joining.

Although limited to four cultures, this study provides direction on the most important features to emphasize in developing interface text and graphics for people with individualist and collectivist values.

FONS TROMPENAARS: CULTURE HITS THE BEST-SELLER LIST

Fons Trompenaars' best seller, *Riding the Waves of Culture*, was first published in 1994; the second edition, with Charles Hampden-Turner as coauthor, came out 4 years later. The book made Trompenaars an international management guru. (According to his web site, he was listed as a top management consultant in 1999, along with Michael Porter, Tom Peters, and Edward de Bono.) His discussion of cultural differences as potential business assets began as a critique of existing Western management theories.

Trompenaars' career has been in industry—working for Shell's Personnel Department in the 1980s and founding his own consulting firm (now Trompenaars Hampden-Turner) in 1989. His practice specializes in cross-cultural mergers, executive coaching, vision statements, and training. Clients fill out surveys on their cultural orientations and cultural competence, and Trompenaars has accumulated a database of over 55,000 responses to support his analysis of culture.

For Trompenaars, management theory is not culture-neutral. Business practices in different countries can look the same but be based on vastly different assumptions:

The issue is not whether a hierarchy in the Netherlands has six levels, as does a similar company in Singapore, but what the hierarchy and those levels mean to the Dutch and Singaporeans. Where the meaning is totally different, for example, a "chain of command" versus "a family," then human-resources policies developed to implement the first will seriously miscommunicate in the latter context. (Trompenaars & Hampden-Turner, 1998, p. 6.)

Culture defines the way each society solves the problems of its social, temporal, and physical environment. Trompenaars uses an analytic framework of seven cultural dimensions: the five pattern variables of Talcott Parsons (1951), which explain how people relate to one another within a social system, and the cultural dimensions for time and the environment from Kluckhohn and Strodtbeck (1961) and Edward T. Hall (1959/1973). The seven dimensions are:

1. *Universalism vs. Particularism*: the extent to which people base their actions on rules or relationships. In universalist societies, people believe it is possible to define universal principles which will be "true" in all situations. Governments are based on the rule of law; economics is a social "science"; and people conduct business as a rational enterprise. In particularistic societies, people temper their actions to fit the social situation. They may support legal principles in theory but bend the truth in a court of law to help a friend; the term *social science* is an oxymoron; and business is a creative art. Universalists base their business dealings on contracts; particularists renegotiate their business relationships when personnel change. Abstract reasoning and statistics persuade universalists; the practical effects of policies persuade particularists.

2. *Individualism vs. Communitarianism*: the extent to which people act on the basis of their own needs or the needs of their social group. Individualists believe that society benefits when all have the freedom to act in their own self-interest. Communitarians believe that society benefits when people discipline themselves and consider the needs of others. Individualists like to work (and play) alone; they expect credit for the personal responsibility they take on. Communitarians prefer to work closely with others; they may vacation in a group; they expect credit to be given to the project team as a whole (not just the team leaders.) Personal incentives (like cash bonuses) motivate individualists; higher-order principles (like quality) also motivate communitarians.

3. *Neutral vs. Expressive Communication Styles*: the extent to which people hide or display their emotions. At a fundamental level, people everywhere feel the same basic emotions (see Gudykunst and Ting-Toomey, 1988), but they do not express them in the same way. People in neutral cultures restrain their feelings; people in affective cul-

tures openly show them. When someone in a neutral culture loses control, their loss of composure is shocking; when someone in an affective culture feels something extreme, their depth of feeling may not be noticeable. When someone in a neutral culture uses irony, the humor is likely to be lost on people who have a passionate communication style; in fact, the ironist may even be thought deceitful. Different cultures separate their emotions from objective, logical decision-making or display their emotions as a legitimate part of the process. In face-to-face communication, each style affects eye contact, rate of speech, tone of voice, facial animation, use of silence, and overall preference for nonverbal or verbal communication.

4. *Specific vs. Diffuse Relationships*: the extent to which relationships with other people are limited or unconditional. People in specific societies may consider coworkers close friends but never see them out of the office; people in diffuse societies become close to fewer people at work but include them in their private life. These different expectations of intimacy reflect culturally different notions of public and private space and of the costs of friendship. Consequently, people in specific cultures often feel free to criticize friends because the friendship doesn't involve a deep level of intimacy. People in diffuse cultures often seem indirect (even evasive) because they know that their words have greater emotional significance. They also take longer to make friends because the cost is unconditional acceptance. Once that relationship has been established, the relative social roles tend to stay the same forever—if you become friendly with a teacher, that person can give you advice even after you establish yourself successfully. This dimension influences both the depth and breadth of emotional involvement and is closely related to communication style. There is also a close link between specific and diffuse relationships and several of Hofstede's dimensions.

5. *Achievement vs. Ascription*: the extent to which status is accorded on the basis of personal performance or social role. Cultures with an achievement orientation value performance and action; doing something is better than doing nothing. Cultures where status is ascribed on the basis of gender, age, family connections, or profession assume that these characteristics are important indications of effectiveness; doing nothing—waiting for opportunities to present themselves—is better than acting in haste. Achievement cultures value youth for its own sake and are often informal; people expect to change jobs and employers during the course of their career. Ascription cultures value experience and maturity; people are loyal to their company and demonstrate respect for the organization by absolute respect for superiors. Networks in achievement cultures are personal; in ascription cultures, they duplicate the official power structure. Titles and degrees are important in as-

cription cultures since everyone needs to know the sources of an individual's status.

6. *Orientation to Time*: the time horizon (past, present, or future) and preference for sequential or synchronic activities. Trompenaars combines the time dimensions of both Kluckhohn and Strodtbeck and Edward T. Hall. People view their actions in historical terms, in the "here and now," or in terms of the future. As a result, they may be resistant to change, follow current trends, or defer immediate gratification for long-term advantage. An orientation to the past or distant future is often found in ascription cultures; an orientation to the present (or near future) is found in achievement ones. Ascription cultures may see time as circular; achievement cultures see it as linear. Different groups also use time in different ways. *Sequential cultures* make schedules and do one thing at a time (Hall's monochronic cultures). *Synchronic cultures* are open to the situation; they let events develop their own momentum and often end up juggling multiple activities at the same time (Hall's polychronic cultures).

7. *Orientation to Nature*: the extent to which people feel they can control, or are controlled by, nature and the social environment. Like Kluckhohn and Strodtbeck, Trompenaars believes that people see themselves as masters of nature or living in harmony with nature (and occasionally mastered by it.) These contrasting positions lead to varying appreciations of human agency and the natural world. People who are *inner-directed* believe they can control external events and are attracted to technology and stories of human achievements; people who are *outer-directed* value nature for its own sake and prefer to work within the status quo. When plans fail, inner-directed people may react aggressively and try again, while outer-directed people bide their time and wait for new opportunities. A control orientation can affect a culture's willingness to change and innovate while an orientation to harmony may promote flexibility and service to others.

Because of the popularity of *Riding the Waves of Culture*, Trompenaars' seven cultural differences are well-known. The framework is very appropriate for intercultural studies but does have one problem—unlike Hofstede's dimensions, it has been applied to (not derived from) a statistical analysis of Trompenaars' survey data.

Trompenaars does not summarize his data by culture; in addition, countries are rank ordered for individual questions but not for each dimension overall. As a result, the distinctions between many of the dimensions are unclear: for instance, one question on neutral or expressive communication style seems similar to another on specific or diffuse relationships. Individualism and collectivism, universalism and particularism, and achievement and ascription all appear related. This makes it difficult to use Trompenaars' set of cultural values experimentally.

Trompenaars as a Guide for Interface Development

Nonetheless, Trompenaars' seven cultural differences make excellent design heuristics—especially when combined with principles of graphic design like the "grammar" of Kress and van Leeuwen (1996). Should the mode of address in messages, text, and graphics be especially polite and reassuring or more direct (and potentially brutal)? How must trust do people need to develop before they invest hours of work in an application? Do they want to be in control or can a wizard help them achieve their goals?

It is easy to speculate on specific effects, particularly for the Web. For instance, sites for people with universalistic values might focus on products, functionality, and price, whereas sites for people with particularistic values might focus on business relationships and service.

Universalism	*Particularism*
Overall focus on products, functionality, and price	Overall focus on business relationships and service
Little information about company	Strong branding
Price comparisons	Customer testimonials
Formal rhetorical style	Familiar rhetorical style (2nd person address)
Sharp focus images that highlight the product	Soft focus images that highlight the customer
Symmetric templates	Asymmetric templates
Consistent interaction style	Use of different interaction models to achieve different goals

In selling to a universalist market, a company like Honda might use the Web to emphasize the innovative engineering used in its new gas-hybrid cars while emphasizing its service and reputation in selling to a particularist market.

Sites for people from individualistic cultures might be designed to enhance individual performance, whereas those for people from communitarian cultures might emphasize group goals and behavior.

Individualism	Communitarianism
Focus on individual self-interest and enhanced performance	Focus on collaborative behavior and group communication
Text reinforces self-image, status, or material success	Text discusses benefits to others
Emphasis on innovation	Emphasis on reliability and quality
Images of individuals using the product (or the product alone)	Images of groups of people using the product (product may never be displayed)
Images of people at play (or working in informal settings)	Images of formally dressed people working with others

When appealing to individualists, a food company like Starbucks might emphasize the rarity, taste, and price of its shade-grown coffee but focus on the benefits to the environment when pitching to communitarians.

Sites targeted at people with a neutral communication style might use a more subdued color palette, reserved rhetorical style, and larger proportion of text to graphics than sites targeted at people with an affective style.

Neutral Communication Style	Expressive Communication Style
Subdued color palette	Bright, expressive colors
Reserved rhetorical style	Exaggerated claims, elaborate words, and emphatic punctuation
More text than graphics	Equal or larger amount of graphics to text
Rhetorical strategy based on objective, logical analysis	Rhetoric anchored in personal stories or narratives
Uncluttered layout	Dense layout
Long- and middle-range images	Close-up images
People in images make reserved eye contact	Strong eye contact or none at all
Use of irony and word play	Use of comic images and animations

A cruise line selling in northern Europe (where most people use a neutral communication style) might show an overall image of one of its boats and emphasize its ports of call. The same line might show the sumptuous buffet (or a collage of its tourist facilities) when selling to a more expressive U.S. audience.

People from specific cultures may prefer web sites that are task-oriented and direct, whereas diffuse cultures may require more information (much of it relational) even for nominally simple web transactions.

Specific Relationships	Diffuse Relationships
Task-oriented information	Information that references the business relationship
Direct rhetorical style	Indirect rhetorical style
Evocation of personal satisfaction	Evocation of public principles
Images of happy, smiling people	Images of reserved, serious people
Specific to general logic	General to specific logic

A U.S. university recruiting students from a specific culture might talk about the starting salaries of its graduates and show groups of happy students (at play or graduation). Students from a diffuse culture would look for information about the prestige of its programs and graduates and be persuaded to come by images of students hard at work on the best new equipment.

People from achievement cultures are likely to be persuaded by sites that help you "do something" (buy a product or recommend a book), whereas people from ascription cultures may require information about who is doing the selling before they will act. It is not that these users don't want to use the Web productively—they do, but first they need to know why they should buy from one organization rather than another.

Status by Achievement	Status by Ascription
Programmatic sites that "do something"	E-commerce functions secondary to organizational information
Focus on outcomes	Focus on reputation and qualifications
Visual angle giving power to the viewer	Visual angle enhancing size of object
Images of youth (both male and female)	Images of older men
Leaders made to appear approachable; names used without titles (but qualifications often given indirectly)	Leaders given full job titles and honorifics
Frequent updates and revisions	Site relatively static (consistent over time)
Many links (broad action hierarchy) presenting many choices	Fewer links (deep action hierarchy) with choices more strongly structured

An established company like IBM would have a built-in advantage over a relatively new company like Dell when selling to people from an ascription culture. It could maximize this advantage by reminding people of its technical leadership and focusing on the CEO. In an achievement culture, such information would be secondary to providing advanced functions (like the ability to build your own PC through the web site). Too much relational information might even be viewed as "information overload" and become detrimental.

Differences in time horizons can also influence the amount of text and the choice of graphic design. Sites with a past orientation need to give a lot of historical context and use traditional color schemes; sites with a present orientation should focus on quick action and use bright, active colors; and sites with a future orientation ought to talk about mission and future plans and consider using a pale (indecisive) palette.

Orientation to Past	Orientation to Present	Orientation to Future
Dark colors with historical connotations	Bright, active colors	Futuristic or dreamy colors
Background information before new information	Current information only	Information on mission and future plans
Emphasis on text and historical information	Emphasis on empowering users to act quickly	Promotion of organizational innovation and use of novel interface features
Images of historical figures or older people	Images of contemporary people doing things	Nonrepresentational or idealized images

These color schemes and rhetorical strategies could also help a site target different domestic audiences, such as conservative investors or early adopters of new technology.

Sequential and synchronic sites might use different interaction models. For instance, a sequential site might use a careful hierarchy to lead people through tasks one step at a time. (To buy a train ticket, a user would first be required to choose a destination. Then the system would provide a list of departure times. Finally, the system would ask for the number of people in the party and let the user choose from available classes of travel.) A synchronic site might ask all those questions on the same page, knowing that some queries could fail and force the user to choose again.

Sequential Use of Time	Synchronic Use of Time
Strongly structured task hierarchies	"One-click" interaction model
Wizards	Tolerance for failure
"Breadcrumb trails" and site maps	Total reliance on search engines

Finally, people from cultures with a control (inner-directed) orientation to nature are likely to celebrate human agency and technology, whereas people from cultures with a harmony (outer-directed) orientation might favor graphics of nature and social cooperation.

Inner-Directed Cultures	Outer-Directed Cultures
Focus on action and technology	Focus on nature and human community
Spare, instructional text	Figurative, metaphoric text
Images of human achievements (e.g., a picture of the Hoover dam)	Images of nature (small-scale images that evoke nature in microcosm or panoramas that emphasize its power)
Bright, active colors	Pale, "natural" colors
Symmetric templates	Asymmetric templates

These variations are all speculative. Unlike Hofstede's work dimensions, Trompenaars' cultural differences have not been extensively used in research, but his work does fill gaps in the analytic frameworks of other cultural values researchers. In addition, five of his seven dimensions apply directly to communication styles and relational goals.

DAVID VICTOR: CAPTURING INTERCULTURAL COMMUNICATION IN A MNEMONIC

One last theorist remains to be discussed. David Victor takes a more historical and categorical approach to intercultural communication, focusing on specific elements of the communication context that can facilitate or hinder mutual understanding.

In his 1992 text, he warned that international business communication is increasingly subject to two contradictory trends: the homogenization of consumer needs (and products) and the intensification of cultural heterogeneity. On the one hand, consumer demographics have converged so that you can buy blue jeans in Japan, Pokemon in Africa, CDs by Ali Farka Toure in the United Kingdom, and Christmas crackers in the United States. On the other, people have begun to resist homogenization by reemphasizing their cultural uniqueness. Strong political-social movements in Quebec, Eastern Germany, and much of the Arab world seek to protect cultural identity. Even Japan, an enthusiastic consumer of Western popular culture, recently instituted a Council on the Japanese Language to investigate the proliferation of loan words in katakana.

Victor promises that anyone can learn to communicate in any culture by using his "culture-general" approach (though they will need to work at it.) A mnemonic—LESCANT—captures the seven most important aspects of international business communication:

1. Language
2. Environment and technology
3. Social organization
4. Contexting
5. Authority conception
6. Nonverbal communication
7. Temporal conception.

The pragmatics of communication tend to be more important to Victor than cultural theories. However, unlike popular books on "Do's and Taboos" (specific things to do, or avoid, when dealing with people from particular countries), his analysis is detailed and comprehensive. For instance, the chapter on language ranks the world's leading languages, discusses the Sapir–Whorf hypothesis, gives examples of linguistic ethnocentrism that can isolate the non-speaker, identifies problems related to linguistic (but not cultural) fluency, lists strategies for working with translators, and cautions against stereotyping based on broken speech.

Unlike most of the theorists, Victor also comments on the role of technological factors in international communication. He describes how a sales opportunity was lost when a business person expected a potential client to have easy access to a phone. Software designers often make the same error, expecting users to have access to the latest hardware and software and the desire to work with the latest release. In many countries, people deliberately choose to work with back-level systems to prevent technical problems and reduce risk.

Victor's discussion of "contexting" (communication that relies on information outside the message) merges Hall's theory with Hofstede's notion of uncertainty avoidance. Communication in high-context cultures is embedded in strong interpersonal relationships: Strangers must develop trust before they can successfully negotiate with one another. By contrast, people in low-context cultures focus on the end product, not the provider. Perhaps as a result, they depend to a greater degree on explicit verbal and written communication, contracts, and commercial law. In addition, people in high-context cultures rely more on nonverbal communication and become skilled in inference. Silence and the settings in which people speak often communicate more meaning than the polite, indirect speech they use. However, this ambiguous style does not create any tolerance for ambiguity or risk.

People in high-context cultures avoid uncertainty by acting conservatively and gathering as much information about their communication partners as possible. This relates back to the emphasis on personal relationships, but it comes with a downside: "Face" becomes central to the communication process. One is forced to protect one's own public persona and the dignity of everyone else. Consequently, people in high-context cultures prefer indirection, harmony, and politeness to argumentation, confrontation, and self-disclosure. Everything works well until low-context people attempt to communicate with them. Then, the differences in communication style may end in shame and mutual loss of face—although the low-context partner often does not know it. (Stella Ting-Toomey [1996]; Ting-Toomey, Oetzel, & Yee-Jung [2001] discussed face negotiation in more detail.)

Victor's wealth of information can be overwhelming. Like the "thick description" of anthropology, it is hard to use the LESCANT system to predict areas of concern because you have to be immersed in the other culture to do the analysis. But if you plan to target just one other culture, Victor's framework is an excellent way to identify areas of agreement and cultural faultlines.

Victor as a Guide for Interface Development

Victor's focus on the pragmatics of business communication has many points of intersection with interface design. For instance, localization often treats the issue of language as a simple technical problem, that is, leave room for expansion and translate the text. But Victor (1992) notes that apparent fluency has its own drawbacks:

> For example, a U.S. businesswoman may represent a company in negotiating with a Japanese firm. The Japanese negotiating team with whom she interacts has little knowledge of English. Recognizing that the woman speaks Japanese, the negotiating team is likely to judge her by Japanese standards. When the U.S. businesswoman breaks Japanese gender-linked social expectations and does not behave as they might expect a Japanese woman to behave, she is less likely to be forgiven than a foreigner unable to speak Japanese might be. (p. 30).

Localized products often evoke similar responses. Users appreciate being communicated within their own language, but untranslated cultural assumptions undermine their good impression. Problematic interface elements include informal text based on different norms of politeness and icons or graphics that violate national and religious conventions. (Paradoxically, interfaces in the original language reduce user expectations and may increase user tolerance for social gaffes.)

Interfaces need to be "culturalized" as well as translated; icons, graphics, rhetoric, and interaction style must also suit the target audience. A German telecommunications company found that investing in product differentiation generated higher acceptance. Pia Honold (1999) ran a series of focus groups and tests on user documentation for cellular phone customers in

Germany and China. German users preferred a thick book identifying all the phone functions; they wanted explicit information in as much detail as possible. Chinese users asked for online documentation; they would lose face if other people saw them looking at a book to solve a problem.

Corporate guidelines for localization should include a LESCANT checklist. The interplay between cultural values and effective communication involves a wide range of pragmatic issues that are easy to overlook.

SUMMARY

So, after discussing all these theorists, what are the most important cultural values to consider when designing software?

As always, "it depends!" If I were adopting just one theorist's model of culture and communication, I would use that of Geert Hofstede or Shalom Schwartz. Hofstede's (1997) work dimensions have a great deal of face validity and are widely used in research, but they are not entirely independent of one another. In addition, his country rankings omit some important nations. By contrast, Schwartz's (1994) cultural dimensions provide an alternative to Hofstede's work and also include rankings. But his seven values are more difficult to grasp and some of the countries included in his list are atypical (Estonia, Bulgaria, etc.) .

If I were picking a set of values from those that show up over and over again in various theories and studies, I would start with these elements of the social environment: (a) individualism and collectivism, (b) power distance, (c) human action (mastery and harmony), and (d) time (particularly the time horizons).

The primacy of the individual or the group seems to be the key consideration for intercultural relationships. Kluckhohn and Strodtbeck, Hall, Condon and Yousef, Hofstede, Triandis, Gudykunst, Trompenaars, and Victor all explicitly reference individualism and collectivism. Schwartz claims to go "beyond individualism/collectivism," but his cultural dimensions reflect the same social dichotomy: The two autonomy dimensions and self-enhancement (individualism) are the obverse of conservatism and self-transcendence (collectivism).

Conceptions of social power (power distance, hierarchy, ascription, and achievement) strongly covary with individualism and collectivism. They should be considered in conjunction with it or folded into the dimension as vertical and horizontal individualism and collectivism (Triandis, 1995). Understanding how humans interact with computers, and with one another across cultures through applications or web sites, also requires an understanding of the ordering of social relationships.

Human action supports a set of themes associated with mastery and harmony. Cultures that support mastery focus on personal empowerment and activity for its own sake, whereas those that support harmony tend to be more passive and perhaps more fatalistic. There is some overlap between mastery, masculinity, and individualism and harmony, femininity, and collectivism, but country rankings do not always bear this out. For instance, Schwartz (1994) ranks France at the bottom of his country list for mastery but Trompenaars and Hampden-Turner (1998) put it near the top of their list for control of one's fate. Nonetheless, the social role accorded technology in different cultures is an important aspect of the human–computer interface.

The fourth dimension is time, though its expression is multifaceted. Cultural views of time show up in decision making predicated on short-term rather than long-term goals, rhetorical strategies that ignore or appeal to history, specific or diffuse relationships, and monochronic or polychronic scheduling of events. However, in general, cultures that focus on the present tend to ignore long-term consequences (whether past or future) and work with other people through more instrumental (specific) relationships.

To round out my selection, I would also pick the two cultural dimensions that most clearly deal with communication style: (a) context and (b) affective (expressive) and neutral communication.

Both overlap aspects of the first four values I mentioned, but more importantly they supplement one another. Considering how people in different cultures show or restrain their emotions enriches Hall's theory of high- and low-context communication. Hall linked individualistic cultures to explicit messages and collectivist cultures to nonverbal messages contextualized within a social setting, but both types of messages can show varying amounts of affect. Most low-context countries, like the Netherlands, use a neutral, unemotional style, but so do some high-context countries, like Japan. Arab culture tends to be expressive and high-context. Affective and neutral communication should be considered in regard to communication context, just as power distance and hierarchy have been folded into individualism and collectivism.

FUTURE DIRECTIONS

Despite the development of virtual teams and multinational business practices, culture is recognized as a far more potent force outside the United States than in it. The expansion of the Common Market, the success of Japanese management (before the collapse of the "bubble economy"), and the globalization of the Chinese and Southeast Asian economies have prompted the rest of the world to reexamine the validity of Western social science and management theories. International journals of intercultural communication, cross-cultural psychology, and management and technology contain work with immediate application to HCI.

Many researchers in China, Japan, Korea, and Southeast Asia are particularly active in identifying alternative cultural paradigms. Michael Bond (in Hong Kong) helped Geert Hofstede identify Chinese values; Tsukasa Nishida, Young Yun Kim, Yuko Matsumoto, Kwangsu Kim, and Taae-Seop Lim have been exploring the validity of other generalizations about communication. Cultural values, with their built-in dichotomies, work well within the European and American academic tradition, but a great deal of innovative research is being done throughout the world to develop more holistic concepts of culture.

Western software development theories are also being subjected to analysis. People like Alvin Yeo have investigated the assumptions underlying design and evaluation practices. Researchers dealing with the transfer of collaborative technologies are finding that basic principles of usability do not have the same utility in all cultures.

In Malaysia, Yeo (2001) discovered that usability tests and other standard methodologies associated with Western software engineering frequently don't work. Malaysian users testing a spreadsheet he developed found it difficult to express their dissatisfaction; because of their collectivist orientation, they censored themselves to preserve his face. Yeo received the most accurate feedback from people already known to him (who knew he needed their negative comments) and from computer experts (whose expertise authorized them to speak critically). He suggests that interface testing in Asia must be based on objective performance measures rather than on subjective user feedback. The International Workshops on Internationalisation of Products and Systems (IWIPS) showcase similar war stories about interface development from people working around the world.

Western bias may even underlie some of our most cherished HCI beliefs. Individualistic forms of decision making and task orientation are not universal. In the late 1980s and early 1990s, group decision support systems (GDSS) were introduced into a number of countries in Asia and Africa. Richard T. Watson, Teck Hua Ho, and K. S. Raman (1994) found that Singaporean groups began with more consensus and showed less change than U.S. groups. They speculated that the need for group harmony undermined individual problem solving. Although the effect was weak—coming after group size, member proximity, and task type—they considered culture a necessary factor in interpreting GDSS outcomes.

Four years later, Gert-Jan De Vreede, Noel Jones, and Rabson J. Mgaya (1998) reported on the introduction of GDSSs in Malawi, Zimbabwe, and Tanzania. They found that incentives for participation were positively associated with endorsement by top management and negatively affected by anonymous input. In these vertically collectivist states, people depended on leaders to make initial judgments. In addition, participants wanted the system to identify the contributions of individuals so that they could decide whose ideas could be developed. Paradoxically, then, successful technology transfer depended on hierarchical influence and claims to "own" ideas, two usability features that the software had been designed in the United States to circumvent. (See Jay F. Nunamaker et al., 1991, for a list of these design goals.)

CONCLUSION

Designing effective interfaces is not just good business. Software for intercultural audiences can bring people together or drive them apart. Inappropriate design that violates cultural expectations perpetuates the digital divide (Shneiderman, 2000, p. 86); software that offends local norms fuels opposition to globalization. The Web is an emerging focal point for conflict. K. Viswanath and Liren Benjamin Zeng (2002) recently documented the role of transnational advertising in anti-globalization protests. Similarly, European privacy advocates have attacked Microsoft for its strategies on personal data.

Interface designers need to think about ways to reduce alienation and cultural bias in software that, by design, cuts across national and cultural boundaries. Incorporating cultural values into our design practices is a good way to start. We also need to test software in a variety of countries and factor culturalization (as well as translation) into product development. Perhaps in the future, software design will be reformed by a new paradigm from Asia, Latin America, or Africa that transcends these problems. Until then, we need to be mindful of the effects of culture and do our best to provide people everywhere with interfaces that support them.

REFERENCES

Barber, W., & Badre, A. (1998). Culturability: The merging of culture and usability. In *Proceedings of the Fourth Conference on Human Factors and the Web*. Retrieved January 24, 2004, from www.research.microsoft.com/users/marycz/hfweb98/barber/index.htm

Benedict, R. (1989). *The chrysanthemum and the sword: Patterns of Japanese culture.* Boston: Houghton Mifflin. (Original work published 1946)

Condon, J. C., & Yousef, F. S. (1975). *An introduction to intercultural communication.* New York: Macmillan.

Cross, S. E., Bacon, P. L., & Morris, M. L. (2000). The relational-interdependent self-construal and relationships. *Journal of Personality and Social Psychology, 78,* 791–808.

de Mooij, M. (2001). Internet and culture. *Proceedings of the Conference on Internet, Economic Growth, and Globalization* (August 8, 2001). Duisburg, Germany: Institute for International and Regional Economic Relations, Gerhard-Mercator University.

De Vreede, G-J., Jones, N., & Mgaya, R. J. (1998). Exploring the application and acceptance of group support systems in Africa. *Journal of Management Information Systems 15*, 197–234.

del Galdo, E. M., & Nielsen, J. (1996). *International user interfaces*. New York: Wiley.

Fernandes, T. (1995). *Global interface design: A guide to designing international user interfaces*. Boston: AP Professional.

Gudykunst, W. B. (1983a). Similarities and differences in perceptions of initial intracultural and intercultural encounters: An exploratory investigation. *The Southern Speech Communication Journal, 49*, 58–65.

Gudykunst, W. B. (1983b). Uncertainty reduction and predictability of behavior in low- and high-context cultures: An exploratory study. *Communication Quarterly, 31*, 49–55.

Gudykunst, W. B. (Ed.). (1986). *Intergroup communication*. London: Edward Arnold.

Gudykunst, W. B. (1987). Cross-cultural comparison. In C. R. Berger & S. H. Chaffee (Eds.), *Handbook of Communication Science* (pp. 847–889). Beverly Hills, CA: Sage.

Gudykunst, W. B. (1988). Uncertainty and anxiety. In Y. Y. Kim & W. B. Gudykunst (Eds.), *Theories in Intercultural Communication* (pp.123–156). Newbury Park, CA: Sage.

Gudykunst, W. B. (1993). Toward a theory of effective interpersonal and intergroup communication: An anxiety/ uncertainty management (AUM) perspective. In R. L. Wiseman & J. Koester (Eds.), *Intercultural Communication Competence* (pp. 16–32). Newbury Park, CA: Sage.

Gudykunst, W. B. (1994). Bridging differences: *Effective intergroup communication* (2nd ed.). Thousand Oaks, CA: Sage.

Gudykunst, W. B. (1995). Anxiety/ uncertainty management (AUM) theory: Current status. In R. L. Wiseman (Ed.), *Intercultural Communication Theory* (pp. 9–58). Thousand Oaks, CA: Sage.

Gudykunst, W. B., Chua, E., & Gray, A. J. (1987). Cultural dissimilarities and uncertainty reduction processes. *Communication Yearbook, 10*, 456–469.

Gudykunst, W. B., & Kim, Y. Y. (1984). *Communicating with strangers: An approach to intercultural communication*. New York: Random House.

Gudykunst, W. B., & Lim, T-S. (1986). A perspective for the study of intergroup communication. In W. B. Gudykunst (Ed.), *Intergroup communication* (pp. 1–9). London: Edward Arnold.

Gudykunst, W. B., Matsumoto, Y., Ting-Toomey, S., Nishida, T., Kim, K., & Heyman, S. (1996). The influence of cultural individualism-collectivism, self construals, and individual values on communication styles across cultures. *Human Communication Research, 22*, 510–543.

Gudykunst, W. B., & Nishida, T. (1986). The influence of cultural variability on perceptions of communication behavior associated with relationship terms. *Human Communication Research, 13*, 147–166.

Gudykunst, W. B., & Ting-Toomey, S. (1988). Culture and affective communication. *American Behavioral Scientist, 31*, 384–400.

Gudykunst, W. B., & Ting-Toomey, S., with Chua, E. (1988). *Culture and interpersonal communication*. Newbury Park, CA: Sage.

Hall, E. T. (1973). *The silent language*. Garden City, NY: Anchor Books. (Original work published 1959)

Hall, E. T. (1976). *Beyond culture*. Garden City, NY: Anchor Press.

Hofstede, G. (1980). *Culture's consequences: International differences in work-related values*. Beverly Hills, CA: Sage.

Hofstede, G. (1997). *Cultures and organizations: Software of the mind* (Rev. ed.). New York: McGraw-Hill.

Hofstede, G., & Bond, M. H. (1988). The Confucius connection: From cultural roots to economic growth. *Organizational Dynamics, 16,* 417–433.

Hoft, N. (1996). Developing a cultural model. In E. M. del Galdo & J. Nielsen (Eds.), *International user interfaces*. New York: Wiley.

Honold, P. (1999). Learning how to use a cellular phone: Comparison between German and Chinese users. *Technical Communication, 46,* 196–205.

Horton, D., & Wohl, R. R. (1956). Mass communication and para-social interaction. *Psychiatry, 19,* 215–229.

Kluckhohn, C. (1951). Values and value orientations in the theory of action: An exploration in definition and classification. In T. Parsons & E. Shils (Eds.), *Toward a general theory of action* (pp. 390–406). Cambridge, MA: Harvard University Press.

Kluckhohn, F. R. (1950). Dominant and substitute profiles of cultural orientations: Their significance for the analysis of social stratification. *Social Forces, 28,* 376–394.

Kluckhohn, F. R., & Strodtbeck, F. L. (1961). *Variations in value orientations*. Westport, CT: Greenwood Press.

Kostelnick, C. (1995). Cultural adaptation and information design: Two contrasting views. *IEEE Transactions on Professional Communication, 38,* 182–196.

Kress, G. R., & van Leeuwen, T. (1996). *Reading images: The grammar of visual design*. London: Routledge.

Lea, M., & Spears, R. (1991). Computer-mediated communication, de-individuation and group decision making. *International Journal of Man-Machine Studies, 39,* 283–310.

Lederer, W. J., & Burdick, E. (1958). *The ugly American*. Greenwich, CT: Fawcett.

Lee, O. (2000). The role of cultural protocol in media choice in a Confucian virtual workplace. *IEEE Transactions on Professional Communication, 13,* 196–200.

Marcus, A. (1993). Human communication issues in advanced UIs. *Communications of the ACM, 36,* 101–109.

Marcus, A. (2001). International and intercultural user interfaces. In C. Stephanidis (Ed.), *User interfaces for all: Concepts, methods, and tools* (pp. 47–63). Mahwah, NJ: Lawrence Erlbaum Associates.

Marcus, A., & Gould, E. W. (2000). Crosscurrents: Cultural dimensions and global web user-interface design. *Interactions, 7,* 32–46.

Markus, H. R., & Kitayama, S. (1991). Culture and the self: Implications for cognition, emotion, and motivation. *Psychological Review, 98,* 224–253.

Nisbett, R. E. (2003). *The geography of thought: How Asians and Westerners think differently ... and why*. New York: Free Press.

Nunamaker, J. F., Dennis, A. R., Valacich, J. S., Vogel, D., & George, J. F. (1991). Electronic meeting systems. *Communications of the ACM, 34,* 40–61.

Ong, W. J. (1975). The writer's audience is always a fiction. *Publications of the Modern Language Association of America, 90,* 9–21.

Parsons, T. (1951). *The Social System*. Glencoe, IL: Free Press.

Pike, K. L. (1954). *Language in relation to a unified theory of the structure of human behavior.* Glendale, CA: Summer Institute of Linguistics.

Rafaeli, S. (1988). Interactivity: From new media to communication. In R. P. Hawkins, J. M. Wiemann, & S. Pingree (Eds.), *Advancing communication science: Merging mass and interpersonal processes* (pp. 110–134). Newbury Park, CA: Sage.

Rafaeli, S., & Sudweeks, F. (1998). Interactivity on the nets. In F. Sudweeks, M. McLaughlin, & S. Rafaeli (Eds.), *Network and netplay: Virtual groups on the internet* (pp. 173–189). Menlo Park, CA: AAAI Press.

Reeves, B., & Nass, C. (1996). *The media equation: How people treat computers, television, and new media like real people and places.* New York: Cambridge University Press.

Said, E. W. (1994). *Orientalism.* New York: Viking Books. (Original work published 1978)

Shneiderman, B. (2000). Universal usability. *Communications of the ACM, 43,* 84–91.

Schwartz, S. H. (1994). Beyond individualism/collectivism: New cultural dimensions of values. In U. Kim, H. C. Triandis, C. Kagitcibasi, S-C. Choi, & G. Yoon (Eds.), *Individualism and collectivism: Theory, method, and applications* (pp. 85–119). Thousand Oaks, CA: Sage.

Sheppard, C., & Scholtz, J. (1999). The effects of cultural markers on web site use. In *Proceedings of the 5th Conference on Human Factors and the Web.* Retrieved January 23, 2004, from http://zing.ncsl.nist.gov/ hfweb/proceedings/sheppard/index.html

Singelis, T. M. (1994). The measurement of independent and interdependent self-construals. *Personality and Social Psychology Bulletin, 20,* 580–591.

Stewart, E. C., & Bennett, M. J. (1991). *American cultural patterns: A cross-cultural perspective* (Rev. ed.). Yarmouth, ME: Intercultural Press.

Stephenson, G. M. (1981). Intergroup bargaining and negotiation. In J. C. Turner & H. Giles (Eds.), *Intergroup behavior* (pp. 168–198). Chicago: University of Chicago Press.

Ting-Toomey, S. (1996). Intercultural conflict styles: A face-negotiation theory. In Y. Y. Kim & W. B. Gudykunst (Eds.), *Theories in intercultural communication* (pp. 213–235). Newbury Park, CA: Sage.

Ting-Toomey, S., Oetzel, J. G., & Yee-Jung, K. (2001). Self-construal types and conflict management styles. *Communication Reports, 14,* 87–104.

Triandis, H. C. (1975). Social psychology and cultural analysis. *Journal for the Theory of Social Behavior, 5,* 81–106.

Triandis, H. C. (1978). Some universals of social behavior. *Personality and Social Psychology Bulletin, 4,* 1–16.

Triandis, H. C. (1980). (Ed.). *Handbook of cross-cultural psychology.* Boston: Allyn & Bacon.

Triandis, H. C. (1983). Dimensions of cultural variation as parameters of organizational theories. *International Studies of Management and Organization, 12,* 139–169.

Triandis, H. C. (1988). Collectivism vs. individualism: A reconceptualization of a basic concept in cross-cultural psychology. In G. K. Verma & C. Bagley (Eds.), *Cross-cultural studies of personality, attitudes, and cognition* (pp. 60–95). London: Macmillan.

Triandis, H. C. (1989). The self and social behavior in differing cultural contexts. *Psychological Review 96,* 506–520.

Triandis, H. C. (1995). *Individualism and collectivism*. Boulder, CO: Westview Press.

Trompenaars, F., & Hampden-Turner, C. (1998). *Riding the waves of culture: Understanding diversity in global business* (2nd ed.). New York: McGraw-Hill.

Victor, D. A. (1992). *International business communication*. New York: HarperCollins.

Viswanath, K., & Zeng, L. B. (2002). Transnational Advertising. In W. B. Gudykunst & B. Mody (Eds.), *Handbook of International and Intercultural Communication* (2nd ed.). Thousand Oaks, CA: Sage.

Walther, J. B. (1996). Computer-mediated communication: Impersonal, interpersonal, and hyperpersonal interaction. *Communication Research, 23*, 3–43.

Walther, J. B., & Burgoon, J. K. (1992). Relational communication in computer-mediated interaction. *Human Communication Research, 19*, 50–88.

Watson, R. T., Ho, T. H., & Raman, K. S. (1994). Culture: A fourth dimension of group support systems. *Communications of the ACM, 37*, 44–55.

Williams, R. (1994). *The non-designer's design book: Design and typographic principles for the visual novice*. Berkeley, CA: Peachpit Press.

Yeo, A. W. (2001). Global-software development lifecycle: An exploratory study. In *Proceedings of the SIGCHI Conference on Human Factors in Computing* (pp. 104–111). New York: ACM Press.

Chapter 5

Managing Multicultural Content in the Global Enterprise

Jorden Woods
Co-Founder, Global Sight Corporation, San Jose, CA

INTRODUCTION AND THE IDEAL ENTERPRISE

This chapter is primarily concerned with the application of multicultural content best practices to a global enterprise. Specifically, the chapter deals with how to apply multicultural best practices not only to a single product or deliverable, but to all consumable content generated and maintained in an international organization on an ongoing basis.

To better understand what the challenges are and how the application of best practices affects processes within a global corporation, I first look at a vision of a fictitious ideal enterprise in the context of multicultural content. To clarify, this ideal enterprise is that organization which is capable of providing every significant customer with the same high quality user experience regardless of country of origin or ethnic background. The easiest way to visualize this ideal enterprise is to view it through the eyes of its customers, partners, and employees on a global basis.

Customer Experience

Every customer experience is clearly a very personal situation that cannot be uniquely determined or parameterized. However, it is possible to generalize a customer's experience given enough interaction in combination with

123

customer profiling and user feedback. In this generalized sense, the ideal experience can be conceived as one in which the customer is provided with or has easy access to useful information and/or functionality that meets the customer's needs, is aligned with the user's preferences, and is provided in a timely and relevant manner.

In this ideal setting, it should be clear that be I a Zulu farmer, a princess in the Thai royal family, or a French nuclear technician I should value my interactions with the company's representatives and its content. In short each interaction will meet my particular needs, be they:

- **Linguistic,** expected or preferred language or dialect.
- **Cultural,** expected or preferred cultural nuance in terms of values, historical allusion, use of colors and icons, relationships, hierarchy, religion, worldview, and so on.
- **Technical,** expected or appropriate weights, measures, protocols and conventions.
- **Economic,** expected or preferred currency, taxation, shipping methods, and so on.
- **Legal,** expected or appropriate statutory or regulatory conventions.
- **Geographical,** expected or appropriate climactic or seasonal norms.

Unfortunately, by not fully understanding their customer's specific needs, global enterprises often unwittingly provide many of their global customers with inappropriate information or functionality in a culturally threatening or linguistically intimidating style.

To further define this ideal experience, consider the following scenario: a U.S. company has launched an existing North American wireless product for the Asian marketplace. Primary markets for launch include China, Japan, Korea, Taiwan, Malaysia, Thailand, Indonesia, and Australia. The list below details the key issues that would need to be taken into account to minimize customer alienation and generate keen user interest:

- Localized product interface.
- Culturally appropriate product size, shape, and color.
- Culturally and linguistically appropriate name and branding.
- Competitive distribution channels and partnerships for product placement, accessibility, and promotion.
- Timely localized information about the product in multiple formats: Web, print, wireless, and so on.
- Country or region specific services which appeal to the primary customer demographic.

- Easily accessible, in-language customer service.
- Adoption of local protocols and standards to maximize interoperability with locally relevant systems, accessories, and infrastructure.

By ensuring that the product meets these requirements in each of its target markets, the customer will feel that they are purchasing a product that appears to have been developed specifically to address his or her needs.

Based on the foregoing analysis, it should now be clear that simply repackaging the device with translations of the original content will be insufficient to generate an ideal customer experience.

Employee or Internal Process Experience

There are many internal process scenarios that can lead to the ideal customer experience outlined in the previous section. For example, employees may be working under tremendous stress in a situation where no processes exist. Such a situation, by the very nature of its inefficiency, is unlikely to generate a return on investment. Instead, I now focus on an ideal environment within a hypothetical global enterprise.

To facilitate the corporate employees in meeting the diverse needs of a global customer base, the ideal enterprise makes information, processes, and resources available for the intelligent generation and update of multicultural information. So, when communicating and interacting with a Zulu farmer, a Thai princess, or a French engineer, appropriate information is available for each customer's particular needs, and efficient processes and resources are available to deliver it quickly and accurately.

More specifically, I examine the particular case of the U.S. company that is intent on launching an existing North American mobile device into the Asian marketplace (China, Japan, Korea, Taiwan, Malaysia, Thailand, Indonesia, and Australia). In this case, the ideal company's marketing and engineering staff would have easy access to the appropriate linguistic, cultural, economic, and technical information to effectively market and modify the product for each target country prior to product launch.

The ideal company's business and information technology (IT) resources ensure the quality and timeliness of the information and functionality delivered. They would do this by developing and providing a process infrastructure for automating and streamlining the tasks dictated by the company's business needs. Furthermore, resources, such as experts in the areas of cultural, legal, and tax accountancy, would be available and would have easy access to the company's process flow.

Therefore, when simultaneously launching a product into markets as different as Indonesia, China, and Australia, the company's employees are able to rely on a breadth of information, process infrastructure, and expertise to rapidly and routinely carry out their tasks. Deadlines are met, expenses meet estimates, the uptake of the product in the target markets is in line with expectations, and the company's brand is enhanced.

This scenario should be contrasted with one in which, information, resources, and processes are not in place. In such a scenario, deadlines are only rarely met, expenses balloon out of control, uptake of product is slow, and the company's brand is eroded or tarnished. Examples of the latter abound in the literature with cases such as the following:

- Nike's initial humbling losses to Adidas in the European marketplace based on a misreading of their "rebel" branding as "hooliganism."
- The failure of the Big Three U.S. automakers to significantly infiltrate Japan's Japanese language-centric, metric-standard, right-side drive road system with English-language, English-system, left-side drive cars.
- The United States' disastrous campaign of feeding the hungry in Afghanistan with food packs the same relative size and color as cluster bombs and with information distributed to a nearly illiterate population without pictures and only in English.
- Walmart's slow and expensive expansion into Latin America in which the U.S.-centric culture did not translate and stores were stocked with electronics based on 110 voltage which would not operate locally.
- The U.S. Milk Council's unforgettable "Got Milk?" campaign, which, for the U.S. Hispanic market, was unsuccessfully conveyed as "Are you Lactating?"

Success, however, can be spectacular as seen in:

- The ability of the Japanese Keiritsu to make rapid progress into the consumer electronics, automotive, and steel building sectors on a global basis.
- SAP's domination of the Enterprise Resource Planning (ERP) space on a global basis from its home base in Germany.
- Nokia's rise from the small country of Finland to dominate the global cell phone marketplace.

With this ideal company as a backdrop, it is now useful to explore the challenges to success in greater detail.

CHALLENGES TO SUCCESSFUL GLOBALIZATION

As the examples in the previous section have clarified, there are multiple challenges to the successful development and management of multicultural information and functionality for global markets. These challenges, however, once analyzed can be seen to fall into three primary categories: process challenges, integration challenges, and technical challenges.

The following sections explain each of these in more detail.

Process Challenges

Historically, companies have globalized their operations organically, with new countries being added on an opportunistic basis through distributor relationships, through acquisitions, and, in some cases, through the development of formal subsidiaries. This approach, by focusing on each country as a separate entity, has typically created organizations that are highly decentralized and only loosely integrated.

As a result, it is not uncommon for each office within an extended enterprise to have different business goals, employ different business processes, and use different technology infrastructure to meet the needs of the local marketplace. Furthermore, it is generally the case that divisions within a single country operate autonomously or with only small consideration for the processes, goals, and technologies employed by other divisions (consider marketing, sales, IT, and engineering, e.g.).

From this analysis one can quickly see that the average global company is not only loosely integrated internationally but is also loosely integrated locally. Due to the Web, there has been more focus recently on cross-functional teams; however, it is more often the case that divisions operate independently or in serial, rather than parallel, fashion.

Today's products and services change continuously, operate under very short deadlines, and demand tight communication and tight integration among all the members involved in the development, promotion, sales, and delivery processes. Under the operating model of a highly decentralized and only loosely integrated global organization, differences between highly incompatible business processes and technology infrastructure are exacerbated by differences in local language, culture, tax law, legal approach, time zones, and governmental regulations.

In addition, business process within individual countries will vary based on the type or purpose of the content that is being localized. For example, the globalization business processes associated with a press release in a single country may be radically different from the globalization business processes associated with a product catalogue or interactive customer support within that same country.

These innate incompatibilities and differences create serious inefficiencies in process and work flow. Leveraging content from one region or country to another can then only be overcome by organizations expending large amounts of both human and capital resources.

The sum of three variables—rate of content change, volume of content, and number of languages or countries—can quickly reach critical values which overwhelm the process and resource infrastructure of the corporation. In short, it can become prohibitively expensive and operationally impossible for a company to maintain its global content so as to meet the high standards of quality and timeliness expected by customers, partners, and employees in the global marketplace.

The greatest casualty for a company succumbing to process bottlenecks is that the company becomes unable to meet its time to market commitments. Once this critical capability is compromised, the company can suffer in the following ways:

- Brand and message fragmentation.
- Distribution of inaccurate, out-of-date, or contradictory information.
- Decreases in customer loyalty due to breach of trust.
- Noticeable functionality lag between applications deployed in different markets.
- A negative return on investment due to globalization initiatives.

Such conditions can therefore hopelessly undermine a globalization initiative and send teams back to the drawing board in search of better solutions. Overcoming process challenges are therefore highly critical to achieving success; however, there are other factors that can halt a globalization initiative in its tracks, and these challenges fall under the category of integration challenges.

Integration Challenges

In a globalization initiative, there are two main points of technology integration which must be tackled to be successful. These areas are (a) interoperability with source content repositories and applications to enable content acquisition, and (b) interoperability with service provider tools and applications (i.e., translators, reviewers, editors). Both revolve around technology interoperability. The first involves interoperability with the technology infrastructure within the enterprise, whereas the second involves interoperability both internal and external to the enterprise with desktop applications.

As Internet technology has matured, IT organizations have deployed greater and more diverse varieties of applications within the enterprise to both manage and benefit from the increasing functionality of this medium. As a result, there has been a proliferation of database- and/or file-driven applications for creating and personalizing content, such as content management systems (CMSs), personalization systems, and application servers.

To facilitate online commerce and to generate interactivity with business processes, companies have also deployed procurement systems, logistics applications, and order processing systems. In addition, other applications have been deployed to interact with legacy systems, such as databases, mainframes, and ERP and MRP systems, as companies have increasingly moved their business operations to an Internet-based environment.

Each of these environments contains content that may need to pass through a globalization process such that it can be made available to an important target customer segment. Overcoming this challenge is therefore the first step in a globalization process. However, integration does not end with the source and target repositories; it is also critical that content provided in the globalization process be compatible with desktop tools used by contributors in the work flow process.

For localization, the tools of choice typically include localized or specialty word processors and translation tools such as Microsoft Word, Trados Workbench, Star Transit, or Déjà Vu. An inability to interoperate with these products may make it impossible to carry out translation and localization of content and therefore will quickly make globalization impossible.

As this analysis illustrates, it is generally the case that the format for content processing is often different from the storage format. As a result, after localization, it is critical that the format be once again modified to enable the storage repository to store and manage the content. This process cycle, therefore, places very significant restrictions on the level of technology interoperability. If interoperability is low, the difference can only be made up through human labor, which is expensive and, more importantly, time consuming.

Technical challenges, however, only begin with interoperability. There are many others that must be considered to maintain the integrity of content during a localization process. These other technical challenges are detailed next.

Technical Challenges

On close inspection, each of the information storage and processing environments within an enterprise contains content in different formats and has different rules or Application Programming Interfaces (APIs) for accessing it. As global content leveraging can only begin by first sourcing content from the organization's content repositories, it is critical that any globalization solution be capable of handling three different sets of parameters simultaneously. These parameters include content language encodings, content storage formats, and application and data repository APIs.

An inability to work with any one of the multitude of permutations of these parameters can lead very rapidly to the demise of any globalization initiative.

As an example, consider first a case of content stored in XML, being sourced from an Oracle database, in English (ASCII-7 language encoding), in the United States and being targeted to an I-mode server, in Japanese, for distribution in Japan. To enable this transaction to occur, there need to be three successful transformations. First, the content needs to be localized into Japanese and then stored in either the Shift-JIS or ISO 2022-JP language encoding. Next, the markup format needs to be transformed from XML to cHTML (condensed HTML), and a template needs to be applied with a different screen format configuration due to the different device requirements. Finally, the content will need to be mapped from the Oracle database in the United States to a Japanese version of SQL-Server in Japan.

Now consider an alternate scenario occurring simultaneously with the process just described: a Chinese HTML page stored in an enterprise CMS being targeted to a WAP server in France, or JSP-based information in a personalization server under a UNIX/Oracle configuration in the United States being targeted to a Brazilian, NT-based ASP server. Obviously, the combinations are enormous and interoperability across all major languages, formats, and applications/repositories is the only road to both current and future success.

Three additional issues now become clear:

1. The systems that are interacting in this example need to support double byte character sets, that is, character encodings such as those for Chinese, Japanese, and Korean.
2. Transformations need to be made on items such as date and time format, naming conventions, and weights and measures.
3. Storage of information within these systems must allow for linguistically correct translation (i.e., no concatenation allowed).

In short, it is critical that these systems be internationalized so as to innately handle these three issues. Without an internationalized infrastructure, there is the high probability for data corruption, loss of content integrity, and the generation of meaningless information.

By conquering all three of these challenges simultaneously, an organization now has the ability to create the foundation for content globalization success. This foundation is explained in greater detail in the next section on best practices.

BEST PRACTICES

Key Terms

I often find that any discussion of multicultural content is best understood once five key terms are explained and agreed on. These terms and their definitions with regard to enterprise content are as follows:

1. **Globalization:** a strategic undertaking whereby an organization creates and enforces standardized business processes on a worldwide basis.
2. **Internationalization:** a lower level strategic undertaking whereby an organization develops and/or deploys people, processes, and systems that perform in ways that are independent of language and culture. This is often thought of as an architectural undertaking that is carried out once and then maintained afterward.
3. **Localization:** a tactical process through which content in one language, culture, and country is interpreted to capture the nuance and conceptual meaning for another language, culture, and country.
4. **Translation:** a tactical process in which content in one language, culture, and country is transformed to have a similar wording in another language, culture, and country.
5. **Personalization:** a strategic initiative through which an organization develops processes to provide specific content that is tailored to appeal to the preferences of a particular individual. On a global basis, personalization begins with language and culture.

The essence of these definitions is that there is a hierarchy to these terms. So, localization is dependent on translation; however, it is greater than it. Globalization and personalization are dependent on internationalization, localization, and translation; however, they are greater than all of them combined. Understanding the relationships inherent in this hierarchy, interestingly enough, lays the foundation for discussing the primary best practices for successful content globalization.

The Three Pillars

Based on my experiences over the last 5 years, three critical best practices can be summed up as the Three Pillars for successful globalization. Each of these needs to be done in succession, and when all three are in place, a stable foundation is formed for a globalization initiative. The Three Pillars are as follows:

I. **Strategy:** Development of a high level documented strategy which has the global buy-in of all significant stakeholders;

II. **Infrastructure:** Development and management of a fully internationalized infrastructure;

III. **Global processes**: Deployment and enforcement of standardized and coordinated content localization processes on a worldwide basis.

Pillar I: Strategy.   A globalization initiative, by its very nature as a worldwide project, is strategic in nature. No matter how small, there are ramifications in multiple countries, in multiple languages and cultures, which affect a company's perception and positioning. As a result, a globalization initiative often becomes a highly politicized event, with multiple groups jockeying for control, visibility, and direction.

As a best practice, setting the strategy is similar to the design phase for an art or engineering project. This stage is critical because without it, the direction and goals of the initiative would otherwise be poorly defined. Without well-defined goals or direction, communication across countries, languages, and functional groups becomes nearly impossible. Where possible, the message will be, at best, confusing and will not enable groups to engage in meaningful discussion.

Instead, the strategy for the initiative should be developed by a working group of key, high-level individuals and then openly discussed as a living document among all significant stakeholders. It is mandatory that the individuals driving the initiative be at as high a level as possible in the organization to ensure that the results of the document be binding and be enforceable throughout the organization.

Conversation among the different groups will then have the chance to stimulate truly global thinking among the participants. Such thinking will, in turn, uncover challenges and lead to the creation of broad-based solutions. The strategy document can then be iterated until a compromise is reached which clearly articulates benefits to the global stakeholder community. Once approved, this vision document will form the basis for the road map for the initiative.

Pillar II: Infrastructure.   Once the business vision for the globalization initiative is in place and blessed by the stakeholders, it is then possible to consider the infrastructure implementation issues for the project. It is at this point that the team queries its infrastructure and decides whether or not it has sufficient and/or appropriate resources, systems, and processes to support the initiative.

Unfortunately, it is all too common that teams develop a business vision and then find out during the implementation process that their goals are unsupportable. Due to the technical nature of electronic media, it is critical that the business vision be coupled with technological due diligence. Any gaps between what is called for in the vision and what is available can then be remedied prior to project initiation.

As an example, consider a team that has decided to roll out multicultural content into the Eastern European and Asian marketplaces. This team decides on its direction, goals, and road map but then neglects to investigate the resourcing and technical aspects of the project. Once underway, the team realizes that it does not know how to support double byte character sets (DBCSs) for the Asian marketplaces or the varied scripts of Eastern Europe. Also, its applications for developing content do not support storage, input, or rendering of the character sets. Furthermore, the team's lack of access to multicultural resources leads unknowingly to the development of template colors and icons that are highly offensive in the target regions.

Such a scenario can be avoided by carefully evaluating the human and technological resource infrastructure to appropriately develop, manage, and update content for the target regions. Thinking through these issues and making changes to the infrastructure prior to project kick-off will then reduce the chances for surprises during the implementation phase, or will make the surprises introduced more manageable.

Pillar III: Global Processes.   Although often overlooked during a planning phase, the life of a globalization project is typically much longer in duration than the initial project development cycle. As a result, preparing for and deploying the infrastructure to support updates and ongoing work is just as critical for success as the initial strategy and infrastructure considerations.

As was noted in the earlier section on challenges, one of the greatest hurdles to any globalization initiative is the process challenge. Not only can this challenge create a massive financial and resource drain, it can nullify all efforts such that no noticeable gains are produced.

The first step to overcoming process-based challenges is to design content localization processes (i.e., work flows, content pre- and postprocessing, etc.) that meet the business needs of the enterprise on global, regional, and local levels. Once this design is completed these processes can then be automated via a standardized infrastructure. The final step, and another key to sustainable globalization success, resides in having visibility to, and the power to enforce, these processes on a worldwide basis.

With global processes in place, the update of content to a worldwide audience becomes routine and, in most cases, predetermined. Crises are greatly reduced, as most situations have already been addressed, and there are mechanisms in place for dealing with problems and surprises. Furthermore, worldwide teams are tightly integrated by the global processes, which means that work is completed quickly, efficiently, and to the satisfaction of all parties involved.

Central Oversight With Local Empowerment (COLE)

In addition to the Three Pillars, the fourth key best practice is known as central oversight with local empowerment (COLE). To understand this concept better in context, it is first instructive to discuss alternative development strategies: centralized and decentralized development.

Centralized. Under a centralized development scenario, a company, division, or group manages all of its content creation, technology, and human resources in one location. Advantages to this approach include the following:

- Resources are coordinated with one voice.
- All business processes receive good visibility.
- Consistency can be enforced.
- Costs and scheduling receive good oversight.
- Single technology infrastructure minimizes technical challenges.

Though appealing because of these strengths, this approach has a significant downside, based primarily on the fact that the centrally developed content is actually to be targeted to other regions. As has been noted, these other countries may have significantly different language, cultural, tax, legal, regulatory, and economic environments that will need to be taken into consideration. If these factors are not taken into account, then the entire centralized initiative can be called into question.

The list below summarizes the weaknesses of the centralized approach:

- Costs for multicultural consultancy can rise exponentially.
- Content may not conform to the needs of countries targeted.
- Scheduling may not meet the needs of country offices.
- Technologies and media offered may not meet standards in target countries.
- Country offices may attempt to redo work created centrally, leading to rising costs.

Centralized development may work in limited cases; however, in the long term this approach usually leads to significant problems. Another approach, which is opposite in philosophy to the centralized approach, is typically called the decentralized approach. It is discussed in more detail next.

Decentralized. Under the decentralized approach, a company, such as a multinational, empowers each of its country offices or subsidiaries to act autonomously to produce content for its local needs. This approach clearly addresses the weaknesses of the centralized approach by putting each region in the driver's seat for its content. As a result, the strengths of this approach include the following:

- Country content is focused on the specific needs of each region.
- Locale-specific issues, such as language, culture, and currency, are addressed automatically.
- Scheduling of content releases meets the needs of each office.
- Promotions for partners, holidays, and local events are locally controlled.
- Content is actively tied into the country offices' sales and marketing plans.

As with the centralized approach, however, there are issues that create problems for the decentralized approach. Most of these problems revolve around the development of severe process and integration challenges that hinder the company's ability to coordinate distributed teams and maintain costs. It should be noted as well, that the greatest challenges arise in an online environment due to the ease with which information from different countries can be accessed and compared. The list below summarizes the problems associated with the decentralized approach:

- Lack of central oversight makes coordination on a regional or global basis nearly impossible.
- Significant incompatibilities may arise in the technological and process infrastructure so that different offices cannot work together effectively.
- Costs skyrocket due to redundancies in resources and technology across offices.
- Global branding, messaging, and look and feel becoming highly fragmented.
- Inconsistencies and contradictory information across countries can create increased customer service and customer loyalty issues.

Interestingly, one can see that the strengths of each approach, centralized and decentralized, cancel out the weaknesses of the other. A hybrid approach, which combines the best of both of these worlds, is known as central oversight with local empowerment. This approach is detailed in the following subsection.

Central Oversight With Local Empowerment

As discussed in the previous sections, the strength of the centralized approach lies in its ability to coordinate and oversee processes with a single voice and technology infrastructure, whereas for the decentralized approach, its strengths lie in enabling local offices to fine-tune content to meet the needs of their locale. The optimal approach would then provide a centralized infrastructure that oversees and coordinates all processes, while all processes would route content through the local offices to enable them to develop and/or review content for their markets.

This approach, known as central oversight with local empowerment, has proven to be one of the most powerful assets to companies such as HP, GE, Xerox, and Dell, all of which have globalized effectively. It should also be clear from this discussion that this approach is fully complementary to the principles of the Three Pillars.

Armed with these best practices, companies now have the knowledge and direction to enable them to develop and manage multicultural content sustainably. The following sections go into greater detail into how these best practices can be implemented within the enterprise to generate financial success.

THE DETAILS OF IMPLEMENTING BEST PRACTICES

So how does an enterprise, division, or group use the concepts of best practices to move itself closer to the vision of an ideal enterprise? The following sections detail the steps a company should take with each of the pillars in order to implement a content globalization initiative with a high probability for success.

Pillar I: Strategy

At the highest levels, the strategy for a globalization project is quite a complicated undertaking, as it incorporates large amounts of financial analysis, risk factors, competitive issues, and so forth. These issues are not in the scope for this chapter or this book, so I assume that the business globalization strategy is in place prior to the development of the content globalization strategy. To simplify the discussion of the content globalization strategy, consider it as moving through three distinct phases: (a) definition of goals and scope, (b) communication and cooperation between stakeholders, and (c) road map and metrics.

Definition of Goals and Scope. The first step in creating a strategy for a content globalization initiative is to synthesize from the business strategy three primary constituents: the goals of the initiative; the breadth of the initiative (i.e., the number of countries and teams included); and the depth of the initiative (i.e., the types of products and content to be included).

As an example, consider the following fictitious plan for a U.S. mobile device company that is planning to increase sales in Asia:

> Problem: XYZ Company has noted that the Asian market is the largest market in the world for mobile electronic devices such as phones and pagers. However, XYZ Company has only a small presence in Japan and Australia and some penetration through distributorships in the other major countries.

> Goal: XYZ Company will increase its market share from current levels (< 1%) to at least 5% in the primary Asian countries within 12 months of product launch. It will carry out an extensive launch and promotion campaign within 3 quarters by selling its key North American products. The initiative should reach breakeven within the first 6 months and then become profitable with a growth rate of 10–15% afterward.

> Breadth of the initiative: Research indicates that the primary Asian markets include the following eight countries: China, Japan, Korea, Taiwan, Malaysia, Thailand, Indonesia, and Australia. Teams from the two offices, Japan and Australia, as well as those in the distributorships will be leveraged. Other offices will be set up over the course of the project.

> Depth of the initiative: Three key products and all their supporting collateral will be launched into the markets. The company home page, product, and support sections as well as two commerce applications will also be globalized to support the launch and ongoing marketing and sales activities.

With the goals, depth, and breadth of the project defined, enough information exists about the initiative to begin to communicate about it with the other country teams. It is at this point that the stakeholders for content globalization can be defined, contacted, and brought into the strategic process.

Communication and Cooperation Between Stakeholders.   Choosing stakeholders is a critical step that can have political ramifications if not carried out wisely. To begin, it is recommended that a senior individual from the group leading the initiative create a committee to represent the globalization project. After the group has been created, it should be officially recognized and given a charter by members of senior management in order to imbue it with legitimacy.

At this point, invitations should be made to international deputies to enable each country office to provide input. Individual country stakeholders should be chosen on the basis of their position and ability to create action in their organization. Alienating key individuals who may be threatening or including individuals with little power is not recommended.

To kick off the strategic session, it is recommended that all important stakeholders be brought together so that they can meet and discuss with each other in person. Such a level of interaction is critical at the kick-off of the project but may not be necessary on an ongoing basis unless the initiative is very large, politically charged, and/or complex. Clearly, however, ongoing interaction and communication is critical to enable the team to work through its issues and develop a plan outlining the recommended strategy.

Time should be set aside to deal with the concerns of each office and develop solutions that are both practical and applicable to as many countries as possible. Should problems persist or should issues escalate beyond the capabilities of the team, it may be necessary to go to outside experts for help and a path forward.

Consider once again, the example of the North American mobile device company launching its products in Asia:

> Consider that the CEO gives the mandate to the Senior Vice President of Worldwide Marketing, Nancy Liu, to carry out the marketing side of the product launch. Nancy would then invite her marketing counterparts in corporate and product marketing, as well as channel partners, distributors, and information services, to be stakeholders in the initiative.

> With a significant company presence in Japan and Australia, these two offices would clearly have input on the marketing and technology aspects, and so several members would be invited from each team. Other stakeholders would be drawn from the one key distributor for the remaining countries plus the operational lead from the team in the new Chinese office.

> Issues to deal with would include a strong belief from the Japanese and Australian offices that they are self-sufficient, not needing any help or input from the United States. The distributor would be interested in charging for additional overhead associated with the launch. Meanwhile, the Chinese team would claim not to have time or resources, but would most likely hire consultants to complete the task.

> Part of Nancy's role would be to push for a strategy and process that incorporate COLE. She would need to work with the offices to have them help each other and buy into a single overarching framework for the project. Additionally, all major cultural, linguistic, legal, technical, and regulatory issues would need to be discussed to ensure that these would be taken into account in the process. Once a solution is reached, the team would document the strategy and put forth a road map for its execution.

Creating metrics and a road map enables the team to measure their performance and schedule out the use of resources. It also enables the team to synchronize their efforts more effectively and work toward common goals.

Road Map and Metrics. A team-developed road map is critical to the success of the initiative because it ensures that the global team is focused on the same goals during a well-defined time frame. Milestones and deadlines keep the project on track and facilitate monitoring of team performance and the overall viability of the initiative. Due to this structure, it is also less likely for team miscommunication to occur as the objectives are well documented and mutually agreed on.

Metrics should be put in place both for the deployment of the initiative, often known as the development phase, and the ongoing operations, or maintenance phase, to provide visibility to the project's success. Different metrics may be of value to different initiatives and organizations; however, there are three metrics that should be applicable in most circumstances:

1. On time/on budget metric—this metric is a measure of how well the team conformed to its projected timeline and budget for initiative deployment.
2. Return on investment (ROI) metric—this metric measures the financial return as a function of the money invested in the initiative.
3. Customer loyalty index or similar—this metric measures customer perception with regards to how well the company's initiatives and behavior coincide with customer desires.

Results from these three metrics allow the team to understand which approaches were successful and which were not and to develop its own specific best practices regarding meeting customer needs, fine-tuning intracompany collaboration, and optimizing automation of its globalization processes.

As an example, consider once again the North American mobile device company launching its products in Asia:

After a number of good international meetings, Nancy Liu is able to understand the dynamics between the different offices and decide on the identities of her key stakeholders. Maintaining a good dialogue between these individuals or teams, she then leads the crafting of an outline of tasks, resources, and responsibilities that will evolve into a road map and milestone chart for the initiative.

Under this dynamic, the team would very quickly realize that they are all important to the success of the project but that they must all compromise if the project is to be implemented. For example, the Australian and Japanese offices may feel that they are self-sufficient but will quickly realize that resourcing needs, project costs, collaboration issues, and interoffice synchronization is best shared with Nancy's team at headquarters and other offices. Meanwhile, Nancy will see that leveraging the other offices as much as

possible will speed time to market and reduce responsibilities and requirements of her team.

Once the road map is laid out, milestones defined, and the metrics put in
place, the teams are now on the same page and can move to the detailed cost-
ing, resourcing, and timing issues. Once these issues are resolved, the team
can now move forward with understanding the limitations and strengths of
their infrastructure.

With a strategy and road map in place, the team is now clear on how it
would like to proceed. Before actual implementing, however, it is critical
that the team evaluate its infrastructure and process to ensure that it can actually handle the demands of its plans.

Pillar II: Internationalized Infrastructure

Internationalization as Best Practice.   Most companies, unfortunately,
are unaware of the true capabilities of their organizations, both good and
bad, with regard to the readiness of their people, systems, and processes to
function internationally. This is a striking assessment given the vast numbers
of global companies; however, it is critical to realize that most global companies are actually a loose federation of separate, local entities.

With this vision in mind, it should be clear that most global entities design products and services for a particular market, such as the United
States, and then ask each additional targeted market to create or customize
similar offerings locally. This process is highly inefficient as little is generally leveraged from the original products or services. As a result, costs for localization are high, time frames for launch or update are long, and quality
can be questionable.

The topic of product localization and internationalization has been covered in depth earlier in this book; however, it is important to note that internationalization and localization do not only cover product design and
coding. Actually, internationalization is an issue that pervades every part of
an organization, from the copy written for a manual to the topics of meetings in boardrooms.

Internationalization is a best practice that, once implemented across
people, systems, and processes, allows a company to collaborate globally
and harness the full potential of its worldwide presence as a competitive advantage. Without it in place, the costs and resourcing necessary to work with
incompatible systems that will corrupt and destroy data can eliminate any
project ROI and make success nearly impossible.

Consider, for example, a very large and respected U.S. company that designs and builds commerce systems based on the EDI standard. Prior to inter-

nationalizing their product infrastructure, they developed a U.S. product and then proceeded to send it to overseas offices to localize in country. After a few releases, the company realized that these had multiplied to over 30 releases in the field, none of which shared the same code base, contained the same feature set, or were well documented internally or externally.

To add to the confusion, each of the company's offices indicated that even after extreme customization few if any of the products were well accepted in their local marketplace and had very low uptake. Suffice it to say that internationalization helped to consolidate all products worldwide onto the same code base and rapidly increased product acceptance and sales in all countries offered.

Implementation Plan.   Once a project has been fully scoped, it is possible to understand if the infrastructure is in place to support its implementation. For example, a checklist should be developed which contains sample questions such as the following:

- Do all systems, processes, and people support the creation, transport, and storage of content for multiple target markets in different languages?
- Is the concept of country or language supported by each component of the system?
- Can the system accommodate delivery of different content to different markets based on user requirements?
- Is all text separated from code for ease of localization?
- Has all concatenation (splicing together of parts of a sentence to create a full sentence) been removed from the infrastructure?
- Can different weights and measures as well as date format be stored and displayed?
- Do authors create content for an international marketplace or a particular country?
- Has the content work flow process been developed to include translation as well as in country content development?

The main purpose of these questions is to uncover areas of the infrastructure that may prove problematic to the implementation of the overall vision. As an example, consider that after the strategy has been developed, Nancy Liu and her support teams carry out an audit on their assets.

After a comprehensive audit, the team discovers the following issues that need to be resolved:

- The operating system for the mobile devices does not support double-byte characters (such as Chinese, Japanese, and Korean).

- Datasheet content has been written in Quark, and it is unknown how to make this information available in Asian languages.
- The Web sites in each country (United States, Japan, Australia, and China) are operating using very different content creation and publishing environments.
- 50% of existing U.S.-based web applications do not support double-byte characters, contain no fields for important Asia information, do not support multiple currencies, dates, and address formats, and use a significant amount of concatenation.
- U.S. product colors and cultural icons are inappropriate for all targeted markets and will have to be reworked.
- The primary channel for content distribution in Japan is I-Mode, a mobile device format for which the company has very limited content.
- There is significant project risk as the team has limited experience with simultaneously launching and maintaining products in different countries with different languages and cultures.
- There is no infrastructure in place for coordinating and updating team members on a worldwide basis with regard to important activities and milestones.

With this information, Nancy and her team are able to create a business plan, requesting budget for the internationalization and localization projects that they need to undertake. They also can create specific Requests for Proposal (RFPs) and submit them to third-party vendors that have expertise in internationalization and localization. Given the team's lack of experience in these areas, the third-party providers should enable the team to hit their aggressive timelines. Subsequently, the team is also able to schedule in realistic times and phasing for the completion of these initiatives prior to product launch.

With a clear understanding of the range of implementation challenges that they need to overcome, the team can now plan and prepare for the ongoing life of the globalization initiative. It cannot be overstressed that the ongoing management of a globalization initiative is the greatest challenge to a team or organization. Planning and implementation have specific deadlines and a limited scope; however, the ongoing project is one of constant change and growth. Managing, adapting, and controlling the ongoing global processes is the ultimate key to success.

Pillar III: Global Processes

Efficient global business content processes typically rely on localization. There are two primary reasons for this. First, people are three to four times more likely to purchase products when they are presented with information that is in their native language (DePalma, 1999). Second, content localization is on average one third the cost of content creation (Drakor, 2001).

Here it is important to note that localization is appropriate where source content can be leveraged into a new market and still be appropriate. So, for example, a global press release regarding a company's record quarterly earnings would be appropriate as it is of global interest. However, information regarding a product that is only available in the United States and has limitations that make it unusable in other countries (i.e., mobile devices that do not conform to local protocols) would not. Additionally, it should be kept in mind that information that does not conform to a consumer's preferred environment (appropriate currency, voltage, cultural attributes, tax and legal system, government regulation, etc.) can make purchase of products and services very difficult or impossible. It is also important to understand that content publishing for global markets may rely on a hybrid approach in which specific information is localized and then combined with locally created content. Reasons for such a hybrid approach could include (a) improved management of a company's global brand and messaging while simultaneously ensuring that information is locally relevant; or (b) better utilization and leverage of enterprise-wide resources.

Consequently, before tackling globalization processes, I first explain conventional content localization processes as they nearly always form the foundation for a globalization initiative.

Conventional Localization Process. Conventional localization process flow includes tasks such as the following (see Fig. 5.1):

1. Change detection on source content repositories.
2. Code and content separation.
3. Content and format normalization.
4. Translation.
5. Translation memory and terminology leveraging.
6. Contextual review.
7. Work flow management and monitoring.
8. Synchronization of content with other locales.
9. Quality assurance.
10. Mapping of localized content to target repositories.

As should be clear from this list, and as has been borne out by extensive analysis, localization costs are generally about three to five times the cost of translation alone.

Each of the steps in the previous list can be a very time-intensive and grueling task if carried out manually under tight deadlines on an ongoing ba-

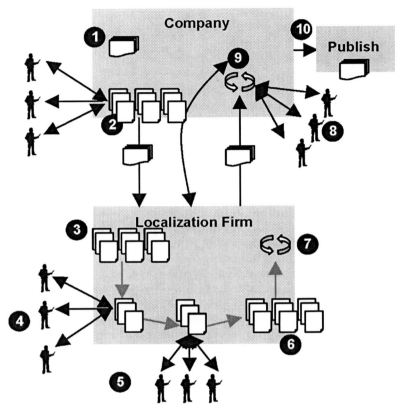

FIG. 5.1.   Localization process.

sis. As volumes increase, timelines tighten, and the number of target markets increases, costs escalate exponentially due to rising resource needs, management overhead, and increased human handling, interaction and environment complexity. If complexity increases to the point where the localization process is too slow to meet deadlines or too inefficient to create relevant content, then a return on investment (ROI) becomes impossible and the investment will never be recouped.

The key to localization process savings is taking situations that are complex and highly leveraged and simplifying them so that they become simultaneously routine and rapid. A combination of process streamlining and automation of the tasks in the list above (see Fig. 5.2) can lead to a 15–25% reduction in costs for localization of new content, and an 80–90% reduction

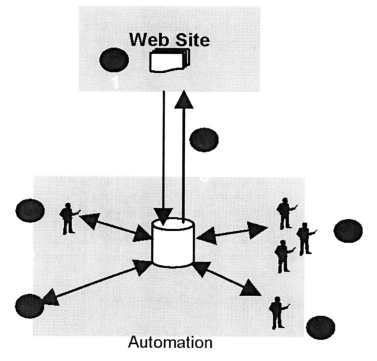

FIG. 5.2.   Streamlined process.

in localization costs associated with content updates. A good rule of thumb is that even for automation of small projects, say 100 pages of content leveraged into three locales with 10% changes per month, cost savings should average hundreds of thousands of dollars on an annual basis.

Once again, consider Nancy's team at the point where they are developing their plans for the ongoing support for the globalization initiative. Specifically, the team must have a plan in place that addresses content, information, and support needs post launch.

Nancy already has understood that decentralized management and operation of the Asian launch will be very expensive and prone to failure. As a result, she has developed a plan based on Central Oversight with Local Empowerment (COLE). Under such a plan, she has centralized her primary technology infrastructure, project management, and project drivers in the United States. Each country office in Asia will be responsible for content review, local content creation, and, in some cases, content translation.

Given the many different types and the large volumes of content that will need to be localized rapidly on an ongoing basis, Nancy explores different technology options for boosting efficiency and quality. To this end, she begins a competitive assessment of localization firms, translation tool providers, and globalization management system (GMS) providers. By choosing an appropriate partner, she hopes to ensure that she can coordinate and leverage her critical corporate information assets on a global basis and consistently hit aggressive localization deadlines.

Nancy also creates a competitive process to review vendors for the management of her source content assets. To this end, her team reviews enterprise content management systems (ECM), portal solutions, and source code control solutions. By combining an appropriate solution for content creation with an infrastructure for global content leverage, Nancy's team can be more confident that they will be building on best practices and generating efficiency and scalability through process automation and streamlining.

The next sections explore in detail the most critical implementation aspects for generating and maintaining success during the ongoing growth and change of a globalization initiative.

WORK FLOW AND PROCESS AUTOMATION

Content change is typically the initiating action in any globalization process, however, there are many other steps that must take place before a globalization process is complete. A typical globalization process can be subdivided into four primary stages: (a) process initiation and content access, (b) content preprocessing, (c) human-based work flow activities, and (d) content postprocessing.

Content Accessibility

Digital content can be stored in one or both of two ways. It can be stored as a file in a file system (e.g., a Word document), as data in a database (e.g., product descriptions as cells in a database table), as a combination of these two approaches (e.g., an ASP, JSP, or Cold Fusion document). As a result, before a document, web page, or specific content string can enter a globalization process, it is imperative that first the content be accessible.

Access can be complicated in multiple ways. The most common ways that content access can be rendered difficult include proprietary storage formats, distributed storage, undocumented or poorly documented storage, proprietary presentation formats, access controls (ACLs), encryption, encoding, and access through other applications. In some cases, companies reduce content accessibility and visibility to improve security; in other cases, it is done to improve content or process management, whereas in still other cases, it may be simply accidental as a result of an implementation approach.

Uncovering your company's or client's particular challenges with regard to content accessibility is critical for the development of any sustainable globalization solution.

Content Preprocessing

Digital content can be acquired in a variety of formats. These can include popular word processing formats, such as Word or WordPerfect; powerful business presentation formats, such as Quark or Framemaker; Web-centric formats, such as HTML, XML, and Javascript; coding formats, such as C++, Perl, and Java; or potentially database formats accessible via SQL.

A key point to note is that any format can be represented as a combination of code and content. The code typically gives instructions regarding presentation and potential functionality whereas the content is a combination of textual information plus graphics.

In a localization process involving nontechnical translators and reviewers, it is generally unlikely that any changes will need to be made to the code. As a result, to streamline a localization process, it is critical that content be separated from the code to simplify the tasks of translation and review.

There is a catch, however, for without the presentation layer or the functionality, the context for the content may be lost, thereby complicating localization tasks. As a result, content to be localized needs to be isolated while at the same time retaining contextual information. This approach maximizes productivity while minimizing the need for rework.

In addition to format, content can be acquired in different language encodings. So, for example, content acquired in the Japanese language will often be encoded as shift-JIS, whereas content in French would be encoded as ISO-8859-1, and content captured in simplified Chinese would be encoded as GB2312. As a result, language translation for digital information leads to a transformation of the encoding as well.

To simplify the processing of language, it is recommended as a best practice that all content to be processed be first converted to Unicode. Unicode is a single encoding for all the world's languages which simplifies language translation by limiting all language processes to a single encoding with well-defined rules. To summarize, there are six steps which must take place to make digital content ready for processing:

1. Separation of code from content.
2. Storage of code for contextual presentation/functionality.
3. Conversion of content to Unicode.
4. Segmentation of content into sentences and/or paragraphs.

5. Separation of content into translatable and nontranslatable segments.
6. Word count for budgeting/timing purposes.

Additionally, it is recommended that all content not only be stored as Unicode but be embedded into an XML-based "container." This container should contain metadata (information about the data) for preserving the integrity and fidelity of the document being processed under a round-trip operation. That is, if I begin with a Word document, I want to ensure that I get a Word document back with the same formatting as the original after localization; the same should be true for a Quark document, an HTML page, or a JSP.

XML is recommended for the container format, as it is a neutral format that is rich enough to preserve the formatting for the most complex documents. To maximize productivity, it is also recommended that all the activities involved with the preprocessing of content be automated as much as possible.

Human-Based Work Flow Activities

Once information has been preprocessed, it is now possible to route it into a human-based work flow for translation, review, and potentially local content additions. Manual work flow processes contain many bottleneck points that lead to high process inefficiency. Automation of work flow processes, on the other hand, can greatly streamline operations leading to much higher throughput, better productivity, and higher quality.

Business processes typically exist within companies; however, they are usually poorly defined or are often reinvented each time they are repeated. For example, a company may publish press releases every week and then distribute versions in 10 different languages across 18 countries, or a company may make daily updates to its online product catalog and localize these updates into the European marketplace. In each case, there are particular initiating activities, which should then be followed by set processes (i.e., distribution into certain languages and marketplaces). Unfortunately, the necessary processes often are not set, and lack of clarity leads to mistakes, missed deadlines, and poor quality.

Automation of work flow processes is not always possible. For automation to be effective, there are a minimum of two prerequisites that should be in place. First, content under a globalization initiative must be categorized or easily categorizable. Second, business rule templates and an infrastructure for initiating business rules must be put in place. Both are necessary, for business rules cannot be initiated without content categorization and content categorization alone cannot lead to automation of processes.

Categorization is simply a fancy term for the type of content, for example, a press release or a product catalog item. With this descriptor in place, it is then possible to associate the content with a particular business process. A codified version of a business process is typically referred to as a business rule. So, for example, consider the press release scenario outlined earlier in more detail.

Press Release Business Process

Every Thursday, an English-language financial press release is created for distribution for the following Tuesday in 10 languages for 18 markets. The original document is created in Quark format and is two pages long. Once the press release is created, it is then sent to five teams in succession, these are:

- An editor who checks the style and formatting and ensures that the content is globally relevant and easily translatable.
- Translators with particular language expertise and expertise in marketing and financial translation then translate the content.
- In country marketing managers then review the press release to ensure that messaging is consistent with their country.
- The in-country legal team then checks the content to ensure that the content will not create any legal exposure.
- Then information is given a final review by an in-country public relations team representative.

Once this process is understood, it can be turned into a standard business rule template that can then be initiated automatically. The business rule can be summarized as follows:

- Every Thursday initiate a work flow process for new content in Quark format in the financial press release folder on a particular server.
- Route this content through five steps: de-Americanize, translate, review, legal, and final review.
- Associate relevant individuals with each step of the work flow process.
- Content must be fully processed by Monday of the following week.

Now, with a combination of content categorization and business rules, the process of globalizing this press release can be carried out automatically on a weekly basis. Obviously, some level of human supervision is necessary; however, all of the manual hand-offs, phone calls, format problems, and second-guessing associated with the fully manual process are no longer necessary.

Content Postprocessing

The final stage of the process is often termed postprocessing. It is in this stage that the content is put into its final form prior to distribution. Postprocessing usually occurs across three axes: format, encoding, and location. So, if as in the previous example, the press release begins as a Quark document, it should complete its localization process also as a Quark document with the same formatting as the original. For print and web documents, there can be significant reformatting work necessary as different languages generally have different linguistic rules and take up different amounts of space on a page. So, for example, German, French, and Japanese translations of English content often show 20–50% horizontal text expansion or text swell, whereas Chinese and Korean typically see horizontal text contraction and vertical text expansion.

Such changes in text size often require different formatting for such items as pagination, table sizes, text boxes, and so on. These changes can be made manually postprocess, or they can be facilitated during document localization during the work flow. In any case, formatting will change between the original language version and the local language version; thus, modifications will need to be made.

Given that the language has gone through a translation process, it is critical that the encoding for the content be appropriate for the language. It should be noted, that though Unicode is a good format for processing multilingual content, it is very rarely a good format for exporting content for distribution.

The reason for this seeming contradiction is that today, very few standard desktop applications use Unicode as a native content format. As a result, Web-based content and that for Quark, Frame, and Word must, in most cases, be converted to national language encodings (NLEs). NLEs are the native encodings that most desktop applications use to render different languages on an electronic device. Common NLEs include Big5 for complex Chinese, KOI8 for Russian (Cyrillic), and Latin-1 for French.

In addition to format and encoding, postprocessing must also specify a storage location for the content. Storage location determination is critical for process optimization because without it, the information will either be lost or could overwrite other important information. Storage location information should specify not only which repository the content is to be stored into, but also a unique address point within this repository that follows some sensible content categorization scheme.

By streamlining and automating the preprocessing, work flow, and postprocessing stages of content globalization, a team can rapidly deliver a return on investment to a globalization project. Another critical factor, which can greatly reduce costs and time and improve quality, is the use of linguistic technologies. The types and benefits of linguistic technologies are covered in detail in the following section.

LINGUISTIC TECHNOLOGIES

Today, there exist three broad classes of linguistic technologies that can be of benefit to content globalization initiatives: (a) translation memory (TM) engines and repositories, (b) terminology management systems, and (c) machine translation (MT) engines.

Each of these technologies has the potential to be employed singly or in tandem to reduce costs, speed time to market, and improve quality. The following subsections explain each technology in more detail.

Translation Memory (TM)

In its simplest form, translation memory, or TM as it is commonly known, is a matching between a sentence and its human-translated equivalent in another language. A TM is often built as a consequence of the translation process, with more "matches" or "source-target pairs" created as translation proceeds. Over time, a repository of such pairs can be mined or translations "remembered" to reduce the cost and time of translation while increasing consistency.

The power of TM relies on the "leveraging" of past translations to reduce the amount of current and future translation. Experience has shown that approximately 20-30% of a single Web site or a large document can contain redundant information. As a result, a TM produced from a single site or document can reduce costs and workloads by at least 20%. A realistic limit for a good TM produced over a few years is that 50% of content can be leveraged directly from the repository. With the addition of a fuzzy TM matching engine, or an engine that can deduce matches based on a percentage similarity, workloads can be reduced by an additional 50%.

Additional factors that can increase the power of a TM include controlled language authoring and the application of the TM to content on the same subject or produced consistently for the same team. Controlled language authoring increases the efficacy of TM by ensuring that content is constructed linguistically the same way each time the same concept is written. In contrast, application of the TM to similar content increases efficacy by ensuring that the concepts to be leveraged are similar from the start.

Interestingly, due to its reliance on human translation, TM is considered by many to be the most reliable and useful linguistic technology that exists today.

Terminology Management

Terminology management technology, or glossary management as it is sometimes known, is similar to TM; however, rather than dealing with sentences or paragraphs, it is concerned solely with words or short phrases. Additionally, terminology is often a linguistic asset that is dictated prior to a translation process rather than being a by-product of it. Terminology is either created by a particular company's marketing department or it is dictated by usage within a particular industry.

To be useful, a terminology management system must be able to do more than aggregate information. As terminology is a by-product of marketing or of an industry, it is important that the teams that use it can edit it and change it over time as terms are added, deleted, or modified. Furthermore, the terminology management system should enable the teams to add sentences to specify context, where appropriate.

Given its nature, terminology management is a critical technology for companies that rely on the precise usage and/or translation of key words. Certain examples where term usage is very important include product names, terminology or jargon for any industry (medical, financial, insurance, technology, etc.), and menu items for applications. In regulated industries, incorrect use of terminology can cost companies millions of dollars in lost time and additional expenses, while in other industries it can lead to confusion and potentially loss of trust and/or revenue.

Integrating terminology management with TM is seen within globalization as a best practice for maximizing consistency and translation quality while simultaneously holding down costs.

Machine Translation

Machine translation (MT) is one of those technologies that appear on the surface to have tremendous potential, but is rarely of significant value in practice. The concept behind MT is that a computing device should be able to "read" a sentence and then translate it correctly into another language based on a complex integration of grammar rules and a cross-language lexicon. Were this theory accurate, the result would be that translation could be done at a rate of hundreds of words per minute for pennies.

Unfortunately, from a technology point of view, language is most often too complex, nuanced, and illogical for today's technology to accurately interpret between languages. As a result, MT is typically relegated to situations in which information is highly technical and very specific in nature, such as in a technical manual, or is of questionable value, such as say in "chat" between consumers on a Web site.

Though it is tempting to implement MT systems, experience has shown that low-cost systems create more problems than they solve, while the costs associated with implementing "high-value" MT systems can exceed the costs for human translation. Though this is the situation today, I would not count MT out entirely, but instead believe that as computing power increases, there will be significant technologies that will be produced in the next few years. These technologies will combine TM, terminology management, and MT to great benefit of the localization and globalization community.

TRANSLATION SERVICE PROVIDERS

Paradoxically, though the skill of speaking multiple languages is nearly universally applauded as a significant accomplishment, good translation is often seriously undervalued by most companies and organizations. Potentially, this seemingly contradictory situation is fostered by the false belief that being able to speak a second language is sufficient for being able to translate between two languages.

Actually, good translation is an art that is a capability of only a small percentage of bilingual speakers. In addition, there are very few translators that can be considered experts across two languages; instead, translators are categorized by their strengths and experience. For example, a particular translator may be highly specialized in translation between German and English for the telecommunications industry, whereas another may be an expert in legal interpretation between French and Spanish.

On the global scene, it is arguable that more wars, misunderstandings, and conflicts have arisen from poor or incorrect translation exacerbated by different cultures than from any other source. As a result, translation should not be taken lightly and should, in most cases, be provided by experts. Keep in mind that your brand, your image, your messaging, and your revenue in another market may rely on the quality of your translators. Also, if you have adopted a model of local content creation in international markets, note that translation is three times less expensive, on average, than native-language authoring.

Translation firms abound in many varieties, from single language pair specialists to huge full service operations employing thousands of people. In determining the appropriate translation firm for your needs, keep in mind four key parameters:

- Translation expertise, including language pairs and specialties (automotive, medical, telecom, etc.).
- Quality and approaches to maintaining quality.
- Cost per word or per project.
- Turn around time.

To keep your choice honest, keep these benchmarks in mind:

- A typical translator can translate 2,000–2,500 words per day.
- A typical reviewer can review approximately 10,000 words per day.

- European languages typically cost US$0.20–0.30/per word.
- Asian languages typically cost US$0.30–0.40/per word.
- A web page costs approximately $100–200 to fully localize.

Also of importance, if quality or multicultural issues are of critical importance, be willing to pay significantly more than the rates quoted here. Note that there are hundreds of stories of companies spending pennies to translate their brand into another market and losing millions of dollars due to poor perception, misunderstanding, and confusion.

OVERVIEW AND CONCLUSIONS

Though the complexity of content globalization is often either underrated (incorrectly believed to be synonymous with poor translation) or overrated (considered impossible), the implementation of best practices to globalization initiatives can enable teams to attain significant return on investment in short time frames.

By intelligently combining planning, an international mindset, global teamwork, and technology, the world's best companies have leveraged their global operations to yield significant competitive advantage. Though the road to success is lined with many challenges, persistence and knowledge can turn content globalization into a cornerstone of your company's long-term sustainability and growth.

Armed with the concepts in this chapter, you are now empowered to walk this road with confidence, avoiding its pitfalls, and attuned to the concepts that will generate success and advancement.

REFERENCES

DePalma, D. (1999). Strategies for Global Sites. Forrester Research Report, Forrester Research.

Drakos, N. (2001). Blueprint for Globalization ROI. Gartner Research Report, Gartner Research.

PART II

DESIGN ISSUES
AND USABILITY ENGINEERING

Chapter 6

Graphics:
The Not Quite Universal Language

William Horton
William Horton Consulting, Inc., Boulder, CO

INTRODUCTION

In today's international market, we must routinely communicate across barriers of language and culture. Even within a single country, products must overcome language barriers. Interfaces for such products, therefore, must also communicate across boundaries of language and culture. So, graphics might seem the logical choice because, as everyone knows, graphics are a universal language. Unfortunately many graphics are *not* universal. Yet, well-chosen and designed graphics can still help user interfaces span cultural and language barriers. This chapter shows how.

What Graphics Are We Talking About?

Graphics in interfaces include icons, emblems, illustrations in online help, and splash graphics. In addition to these interface graphics, the principles discussed in this chapter will also apply to the *overall* display, which is perceived as a whole graphic before being parsed into separate fields and buttons. The principles may also apply to subject matter content, for example, illustrations of a product or a diagram in an e-learning lesson.

Graphics Are Not Universal?

The widespread use of wordless road signs in Europe and the proliferation of icons in graphical user interfaces has given many designers the notion that graphics speak a universal language. Although many graphics do work for more than one culture and can sometimes communicate across boundaries of language, graphics are subject to cultural interpretation (Forslund, 1996). Without careful choices and appropriate rendering, expressing ideas graphically is no guarantee against misinterpretation when viewed by someone from a different culture.

WHY USE GRAPHICS IF THEY ARE NOT UNIVERSAL?

No, graphics cannot totally replace words. However, with careful design, they can enhance the effectiveness of interlingual and intercultural user interfaces. Well-designed graphics can:

- **Reduce the number of versions of a product.** By reducing the level of language skills required, graphics can make an interface usable by second-language users.
- **Reduce translation costs.** Communicate as much as possible in graphics. Even though cultural differences may prevent us from saying everything in graphics, we can still lessen the amount of text that must be translated, lessen dependence on verbal language, and lower the possibility for erroneous translation (Jones et al., 1992).
- **Ease learning.** It is easier to see and understand than to read, translate, and then understand because visual images are less ambiguous and more memorable than equivalent text.
- **Improve comprehension.** Those who read in a second language rely more heavily on graphics than those who read in their first language because the understanding imparted by the graphic helps readers translate the text (Rochester, 1992).
- **Take advantage of an already existing body of recognizable symbols.** Thanks to global communications in the forms of movies, the Internet, and television, there is a greater reservoir of common images from which we can select, further reducing the need for text.

SELECT AN APPROACH

Two approaches of designing products for international markets have emerged: globalization and localization (see Fig. 6.1.) *Globalization* seeks to make products general enough to work everywhere and *localization* seeks to create custom versions for each locale. Some view globalization and local-

Globalization   **Localization**

FIG. 6.1. Two approaches of designing products for international markets.

ization as opposite extremes (Kostelnick, 1995). Each approach can be applied to graphics, but neither is sufficient by itself.

Globalization Approach

The globalization approach seeks to "bridge rather than defer to cultural boundaries" (Kostelnick, 1995, p. 182). Also known as *internationalization*, globalization is "creating a product and its documentation in such a way that it makes no assumptions about human language, local customs, or character set. The resulting product is location neutral" (Kumhyr, Merrill, Spalink, & Taylor, 1994). Globalization foregoes assumptions of a single language or locale (Kano, 1995). The aims of globalization are closely aligned with those of universal design, "the design of products and environments to be usable by all people, to the greatest extent possible, without the need for adaptation or specialized design" (Center for Universal Design, 1997). Globalization of graphics has a standing tradition reaching back to the ISOTYPE symbol system of Otto Neurath (Neurath, 1936) and the charting figures developed by Rudolf Modley (Modley, 1937).

However, globalization alone is seldom sufficient for products marketed to more than a few similar cultures. The concepts that must be depicted in user interfaces are complex and often novel. It is arrogant of us to think we can create images that will have the same precise meaning to everyone everywhere.

Localization Approach

Localization is "making a product specific to a locale" (Kumhyr et al., 1994). The localization approach maintains that graphics are a language learned through experiences within a specific culture (Kostelnick, 1995). And, users in different locales are fundamentally different and require a highly customized user interface (Fernandes, 1996). Advocates of the localization approach point out that a person's culture affects how that person interprets graphics (Hagen & Jones, 1978). Users are much more likely to recognize images of common objects that resemble the ones they see every day (Fuglesang, 1982).

Cultural misinterpretations of supposedly universal graphics have plagued international commerce, politics, and travel for centuries. Studies of

African peoples unfamiliar with pictorial methods found that both children and adults required training to recognize familiar objects shown in photographs and line drawings, and they perceived depth only with difficulty and only when overlap was present (Deregowski, 1974). On the other hand, Australian Aborigines objected to overlap used to show depth and perspective in pictures of birds because the overlap hid a wing or foot (Gombrich, 1969). Likewise, many people in Western cultures don't "understand" modern art or assume the perspective of ancient Egyptian art is due to unskilled drawing.

Localization alone, however, is not a practical solution. Culture is too complex and there are simply too many subcultures within subcultures to produce versions localized for each. Culture is not limited to nationality, language, and religion. It includes age (youth culture, Cold-War generation), employment (IBM culture, the HP Way), gender, caste and social class, wealth, level of education, and even degree of Internet socialization. What "culture" is a 28-year-old single mother who was born and raised in Germany, raised up as a devout Catholic, attended MIT, and now works for the Hong Kong office of a Swedish telecom that conducts business in British English? The answer is that she participates in multiple cultures. And she, rather than the statistically average national stereotype, is becoming the norm.

Culture is a somewhat brutal abstraction. Hundreds of millions of people do not form a monolithic culture. Cultural stereotypes true in the abstract are often wrong in the particulars (Hao, 1999) or unimportant (Mrazek & Baldacchini, 1997). People read graphics; cultures do not. Though informed by cultural experiences and meanings, interpretations are not bound by cultural traditions. Taking into account all cultural factors would require thousands of distinct versions for every product.

A More Practical Approach

While contrasting globalization and localization may enliven a faculty cocktail discussion of postmodern artistic sensibilities, such contrasts do little to guide user interface designers with limited budget and time with which to make their products usable around the globe. The problem is that depicting these two approaches as diametrically opposed belies their synergy in overcoming cultural barriers. Designers are redrawing the diagram to reflect a more pragmatic strategy (see Fig. 6.2.) The strategy is simple. First, globalize to make the graphics work for as many people as possible. Then, localize the graphics that are not truly universal to specific cultures.

FIRST, GLOBALIZE

For pragmatic reasons, the first step in designing graphics for an international product's user interface is usually to globalize graphics, that is, to

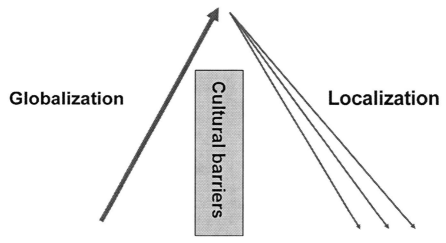

FIG. 6.2.   Globalization and localization work together to create products for international markets.

make them understandable by as many different cultures as possible. Globalization reduces the number of separate localization efforts required and cuts the number of individual graphics that must be localized.

The core strategy for globalizing graphics involves three concurrent tactics: base images on globally common experiences, draw graphics so everyone will recognize them, and eliminate culture-specific symbols.

Pick Globally Recognizable Images

The form of globalization with the highest return on investment is picking globally recognizable images. These images rarely require translation and seldom offend. But picking globally recognizable images requires thinking about ideas common around the world—or at least common among users of your product. It may also require sensitivity in how objects are drawn.

Show Literal Meanings

Where possible, use graphics for their literal rather than symbolic meanings. Showing a hand performing an instructed procedure is a lot less risky than showing a hand gesture signaling that the procedure was completed correctly. The vast majority of misinterpretations of graphics occur in the symbolic realm, so, to minimize such misinterpretations, avoid symbolic uses of graphics and clearly indicate in style and labeling that your graphics are meant to be interpreted factually.

Base on Common Imagery

For a palette of widely understood imagery, draw on science, mathematics, TV news, movies, and international sports. All of these endeavors can suggest symbols and examples that are relatively free of political and religious connotations. Table 6.1 shows some examples.

Use Generic Objects in Examples

Replace examples that are specific to your locale and culture with ones that more people can recognize and understand. For example, if you are talking about money, instead of showing the currency of just one country, show a mixture of currencies (see Figs. 6.3 and 6.4.)

If you need an arbitrary object for an example, avoid ones associated with a particular nation or culture, as shown in Fig. 6.5. Instead, substitute an abstract shape or a neutral object perhaps.

Horton's Rule of Global Recognizability

People who use computers watch movies. If an object appears frequently in Hollywood blockbusters of the 1930s through the 1980s, it is a good bet most users will recognize it, even if it is not common in their culture. Though not ideal, such an object will probably work.

Generalize Images

Even if we base our graphics on global ideas, we must take care to draw them in a way that makes them globally suitable and recognizable.

To avoid needless repitition of advice found in other chapters and to forego insulting your intelligence, we are not going to repeatedly tell you that you need to:

- Research target cultures thoroughly.
- Test graphics with target cultures.
- Involve persons from target cultures in design.

Consider these things said.

TABLE 6.1

Common Imagery

Space exploration		Space Shuttle launches, astronauts and cosmonauts floating free in space, and photographs from the Hubble telescope have been seen around the world.
Medicine		Medical doctors everywhere prescribe pills, use hypodermic syringes to administer injections, examine X-rays, and wear stethoscopes around their necks.
Mathematics		For those in scientific and technical disciplines, mathematical symbols used for their conventional meanings can communicate clearly. The plus and equals sign work for just about everybody. More advanced symbols like the sigma for summation, and delta for change require a bit more knowledge. The minus sign, though, can be confused with a hyphen.
Transportation		You can safely assume that people who use computers are generally familiar with trains, planes, busses, boats, bicycles, motorcycles, and automobiles.
Sports		The Olympics mesmerize most nations, Formula 1 auto racing covers five continents, and the World Cup may draw a television audience of over a billion. Avoid local forms such as American football.
Business		Similarities unite business offices around the world. The desktop computer in Stockholm used to send an e-mail message looks a lot like the one in Sao Paulo that receives it. Desks, chairs, clocks, fax machines, and other office equipment and devices provide a common basis for communication.
Consumer goods		Products, such as cameras, camcorders, binoculars, cassette recorders, film, batteries, light bulbs, and eyeglasses, are marketed throughout the world. Their images are likely to be familiar to computer users.

FIG. 6.3.   Before.

FIG. 6.4.   After.

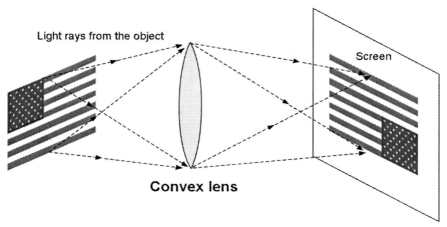

FIG. 6.5. Avoid cultural or national imagery.

Draw People With Whom the User Can Identify

Detailed and realistic images of people inevitably carry cultural and racial identifiers. Therefore, depicting people in interfaces presents special problems. Images of people carry all the cultural characteristics of the people shown. Hewlett-Packard confronted this situation when they created an icon for a software agent, a program that automates separate activities in your computer. It depicted a black-haired man in a white shirt and tie wearing dark sunglasses. The sunglasses were, perhaps, to suggest a secret agent. I heard a woman from Japan refer to it as the "hoodlum" icon. A New Yorker called it the "California Joe Cool" icon. Consider the issues raised by the icons in Table 6.2.

TABLE 6.2

Problematic People

	Users had to click on this button to run a computer program. Many users thought it concerned a person of a particular race, gender, age, and economic class.
	Is this an emblem of data security or of totalitarian oppression?

Nothing represents a specific culture more than an image of a specific person. If you must use an image of a person:

Dress People Modestly. Use common, conservative business attire as your model. Avoid loud patterns, bright colors, or overly casual wear. Although in Western Europe nudes are readily accepted in TV and print advertisements, nudes are rarely acceptable in the United States (Ogilvy, 1985), and in much of the Islamic world only women's hands and eyes may be shown.

Minimize Signs of Economic and Social Class. Dress people in simple business attire; avoid accessories that imply wealth or position, such as jewelry, furs, high fashion, exotic cars, and so on. Especially, avoid emblems of religious value.

Keep Relationships Between People Simple and Businesslike. Show people interacting in a polite and professional way. Be careful about implications of status and power, which in many cultures depend on age and gender as much as on job assignment.

Draw Generic Hands. To show an operation requiring the use of a hand, draw the hand as generically as possible. Minimize gender and racial identifiers, or even use a cartoon hand (Jones et al., 1992; see Fig. 6.6.

Focus attention to the task the work is performing rather than on the person it is possibly attached to (see Fig. 6.7). If only one hand is to be shown, make it the right hand, because in many Arabic countries the left hand is reserved for toilet tasks.

Read more about using hands as symbols later in this chapter.

Avoid Cultural Stereotypes and Overt Gender Roles. Avoid linking activities with people of any particular gender, race, or age. Cultures differ in the roles women and men play and the jobs they can perform. In much of

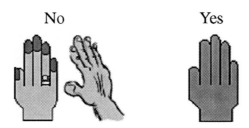

FIG. 6.6.   Draw generic hands.

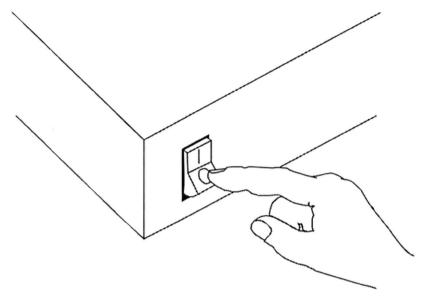

FIG. 6.7.   Show a right hand performing tasks rather than a left hand.

Western Europe, Australia, Canada, Latin America, and the United States, women are found at all levels and in all departments of business, working alongside their male colleagues. This trend, although relatively recent, is so pervasive that a graphic depicting an all-male group of computer users would be thought of as odd or dated.

This equality is not universal, though. Throughout the world, the jobs and roles of men and women differ widely. Even within a single nation or culture, we find extremes. In some Islamic countries, women play little or no role outside the home. In still others, women can fill technical and professional jobs, but cannot work in direct contact with men in the same roles. In Pakistan, however, a woman has served as Prime Minister (Grove, 1989).

The Swedish furniture manufacturer, IKEA, adopted very different graphics for their home page, depending on country (Rahmat, 1999). The home page for Italy showed a couple reclining in an amorous embrace on a sofa. The woman's scarlet dress had slid well above mid thigh. The equivalent page for Saudi Arabia showed a father, dressed in traditional Saudi robes, pushing a shopping cart in which rode his son, similarly attired.

Take Care With Body Language.   Whenever you show people, you tell a story. The way they stand, the way they sit, the arrangement of hands and torso provide powerful cues as to status, respect, and power. Unfortunately, cultures differ in how they interpret these cues. Furthermore, relationships among coworkers and between workers and supervisors vary greatly around the globe (Boiarsky, 1992). Take the simple act of crossing your legs. A well-known Thai journalist and political reformer was ejected from Parliament for sitting with her legs crossed (Axtel, 1991). Princess Diana made news by sitting in public with her legs crossed at the knee, rather than at the ankle as is always done by Queen Elizabeth II (Axtel, 1991). Many American men consider the practice of crossing the legs at the knee effeminate although it is common among European men. Other cultures keep their feet firmly on the floor (Morris, 1977).

Try Cartoons Where Appropriate. Use cartoons, simplified line drawings, or stick figures to depict people so as to de-emphasize gender or race. Do not design cartoon characters for humor but for generality. Remember, humor is quite cultural, and many people have no sense of humor whatsoever (Jones et al., 1992).

Even better than using a person, create cartoon characters by animating objects from the reader's world of work (as shown in Fig. 6.8). Pencils, calculators, computer terminals, coffee cups, filing cabinets, copying machines, and even technical manuals are good subjects (Kittendorf, 1981). Keep in mind the purpose of your graphic, however. Make dangerous things look fierce and menacing; make helpful things appear safe and likable (Frye, 1981).

Be sure not to overdo the humorous aspect of the cartoon. A slight touch of whimsy is better than striving for the belly laugh.

> **Guideline for using people in graphics:** Show people only where necessary and represent them by generic figures rather than realistic photographs or drawings.

FIG. 6.8. Animate objects from the reader's world of work.

Make the Meaning of Position Explicit

Values can be associated with position or direction. For example, righteousness and power are associated with the right hand in Western cultures. In the Chinese tradition, however, honor is on the left hand and on the right is considered violent and self-destructive (Cooper, 1978). Where could this concept become an issue? Consider a web page comparing your product with your competitor's. Where would you place your competition's product in relation to your own to ensure a favorable effect?

Suppress Unimportant Details

Details that could inform one audience may confuse or distract another. For international symbols, design objects "sufficiently abstract that the audience does not have a preconception of their meaning" (Grove, 1989, p. 141).

There is a fine line between making an image recognizable and making it culturally specific. Pick graphics your international readers can identify, but take care to include only those details that enhance recognition. Some suggestions are:

- **Omit details you do not need to show.** Do you need to show a specific power plug, or will most users already know what a power plug looks like?
- **Disguise or diminish national differences**, like clothing styles or decorative product details. Choose a viewpoint that omits such details, delete the details, or selectively blur them.
- **Use an icon or simplified drawing** instead of a realistic drawing or photograph.
- **Obscure or omit textual labels on devices.** For instance, show keyboards with blank keys. Indicate particular function keys by position, not by name or label.
- **Show all possible instances** if you cannot disguise variable features.

Show the Best Known Version

Many common objects come in multiple versions. Forego parochialism and use the version that is most common throughout the world. For instance, draw a measurement ruler with 5 or 10 divisions per unit instead of 4 or 12. The divisions can thus represent millimeters or tenths of an inch. For an automobile, select a small boxy model that is sold in many different countries. For a truck, substitute a delivery truck for a pickup truck, which is not as common in Europe as in America (see Fig. 6.9.)

FIG. 6.9.   Show the best known version.

If you cannot find an international version, then pick the version familiar to the largest number of users.

Mix Several Versions

If there is no universal symbol for a concept, you may have to create several different versions of the graphic and use a different one for each market. Another approach, however, is to combine the several versions into one. The plurality of images produces a symbol more general than any one symbol alone. Even if users do not find their own version represented, they may still recognize the concept in what the separate images have in common (e.g., see Fig. 6.10).

Make Reading Direction Explicit or Unimportant

The user's accustomed reading direction affects the direction in which they scan a graphic and the order in which they read a sequence of graphics. In the Middle East, an advertisement for laundry detergent showed dirty clothes on the left, a box of the detergent in the middle, and clean clothes on the right (Hartshorn, 1989). Why did it fail? It failed because Arabic is read primarily right to left.

It is not always easy determining reading direction. Even languages like Arabic and Hebrew that are read left to right reverse direction for proper nouns and terms imported from left-to-right languages (Merrill & Shanoski, 1992).

Currency

Users

Religion

FIG. 6.10.   Mix several versions.

If you must arrange objects left to right, show direction with arrows, as shown here in Fig. 6.11.

To imply sequence, stack items vertically, as shown in Fig. 6.12. All languages can be read from top to bottom on a page.

A General Approach to Generalizing Graphics

I suggest a three-part approach to meet diverse expectations, tastes, and learning strategies:

1. **Avoid extremes.** Moderation minimizes the confusion and offense any graphic may offer. Strive for a balance between formality and informality. It might be as simple as rounding the edges of a shape, but maintaining the precision of the lines. (See Fig. 6.13.)

2. **Maintain neutrality.** Graphics are inherently more neutral in tone and emotion than text. Avoid techniques that give the graphic a definite personality (Rochester, 1992). For instance, some users may find the graphics shown in Fig. 6.14 too cute:

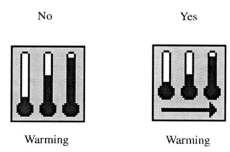

FIG. 6.11.   Show direction with arrows.

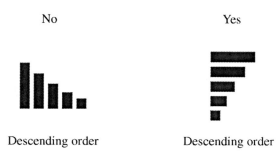

FIG. 6.12.   Imply a sequence by stacking vertically.

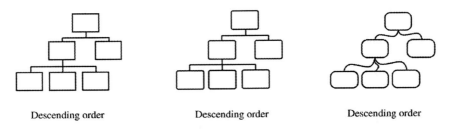

Descending order   Descending order   Descending order

FIG. 6.13.   Avoid extremes.

FIG. 6.14.   Maintain neutrality.

3. **Make the graphic multipurpose.** Design graphics so that they can be read in different ways for different purposes. An example (see Fig. 6.15) is a map with numbered items along a suggested route. Those who want a step-by-step procedure can follow the route; but, those who prefer to learn on their own can explore. Conceptual overviews and maps that

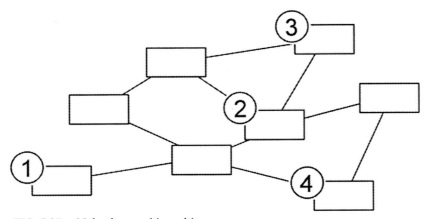

FIG. 6.15.   Make the graphic multipurpose.

illustrate the general hierarchy or organization of the system or the document are especially helpful (Spragins, 1992). Keep in mind, though, that some cultures use different methods of categorization or may even lack a strong concept of categorization (Halio, 1992).

Replace Culture-Specific Symbols

The symbols we may use to encode meaning or to decorate a graphic can have vastly different associations in different cultures. We must ensure that the different associations do not contradict our intended meaning. Here are some specific problem areas.

Puns and Verbal Analogies

Verbal puns rely on objects whose names sound like that of the concept. Table 6.3 shows some examples.

Because puns require a subtle knowledge of the language, those for whom this language is not near native will rarely get the meaning. Puns do *not* translate well.

TABLE 6.3
Avoid Puns and Verbal Analogies

Image	Intended meaning	Problems
	Mouse in a computer system	In some languages the name for the box that controls the screen pointer is not the same as that of a small rodent
	Posting data in a database	This image only works in English where the word post has the additional meaning of mailing a package or letter.
	Scale of a map	In many languages, the name of a balance scale and the relationship between distances in the world and a map are not the same.

Mythological and Religious Symbols

We do not all share the same religious and mythological symbols. Table 6.4 shows examples of symbols to avoid.

Question any symbol that has religious associations. Such symbols can slip in quite innocently: A charting program may use six-pointed stars and crosses when plotting data points; a crescent might seem a good symbol for nighttime, yet it is a common emblem for Islam.

Totems

From prehistoric times, we have used animals as totems, that is, as symbols of properties we feel these animals possess, like courage, intelligence, speed. Consider how many governments and sports teams use the eagle or lion as an emblem. The problem with using animals as symbols is that we do not all agree on which of its characteristics the animal represents, as shown in Table 6.5.

TABLE 6.4
Avoid Mythological and Religious Symbols

Image	Intended meaning	Problems
	Fatal error	Would someone raised a Hindu, Buddhist, or Zoroastrian recognize the grim reaper?
	Minor glitch	Some may not recognize this as a gremlin, but see instead a bat with a long tail. A gremlin can be a symbol of a minor problem or a monstrous evil.
	Assistance or help in solving problems	Only 13% of Japanese users of a product recognized the Red Cross emblem of the intended meaning (Kumhyr et al., 1994). In Islamic countries, the local version of the Red Cross is known as the Red Crescent.

TABLE 6.5

Avoid Totems

Image	Intended meaning	Problems
	Loyalty Search, retrieve	To Americans, dogs are pets or hunting companions; to many Asians, they are food. In come cultures, dogs are considered unclean or unholy.
	Wisdom Expert system Training	In the Greco-Roman tradition, the owl represented wisdom and intelligence and was the symbol for the goddess Athena. In southeastern Asia, the owl is considered a particularly brutal and stupid bird.
	Savings account	In the U.S., the piggy bank is associated with thriftiness. However, to Muslims and many Jews, the pig is considered unclean and unholy. An advertisement for a refrigerator failed in the Middle East despite the fact that its compartment clearly had room for a large ham (Kumhyr et al., 1994).

National Emblems

Do not casually diminish the flag, currency, coat of arms, or other emblems of a country. This can often happen quite innocently. Imagine a symbol of European economic unity that shows a businessman wearing a suit made of the flags of the European countries. Somebody is the lapels and somebody else the seat of the pants. If you must use such a national emblem, make sure you show it in the correct orientation and drawn to the correct proportions.

Colors

Because the symbolic meaning of color varies from culture to culture, ignorance about color associations may cause miscommunication (Cooper, 1978; Dreyfuss, 1984; Thorell & Smith, 1990; White, 1990). Designers for the U.S. Indian Service, for instance, inadvertently biased voting in a Navajo election by using color codes for the candidates. To the Navajo, colors are ranked, with blue being good and red bad (Hall, 1959). According to Jill Morton (www.colormatters.com), in Catholic Europe, purple symbolizes the crucifixion and death of Christ whereas in the Middle East, it symbolizes prostitution. Yellow, which signifies caution and cowardice in Western European cultures, may represent grace and nobility in Japan. In the United Kingdom, first place is often awarded a red ribbon, not blue as is common in the United States (Jones et al., 1992). Clearly, there is not global consensus on the meaning of colors.

Does this mean you should never use colors in international user interfaces? No, color can prove especially valuable, such as making logical categories explicit. If you need color, follow these guidelines:

- Make the design work in black and white first. Then add color to make it work better.
- Use color only in technical and business contexts so that the context does not trigger a symbolic interpretation.
- Clearly define your color scheme and make color codes explicit.

For more explicit advice on using color, see chapter ?

Hand Gestures

Almost every configuration of the human is an obscene, rude, or impolite gesture somewhere in this world (Axtel, 1991). Especially avoid clenched fists and projecting fingers. Table 6.6 shows some examples.

Body Parts

Parts of the body have symbolic meanings and local associations that can thwart their use as symbols. Table 6.7 shows some examples.

Remove or Translate Text

Icons and pictures that depend on text for their meanings require redesign or translation before they will work for users who speak another language. If possible, redesign these graphics; if not, flag them for translation.

Words and Abbreviations

Design images free of text, especially abbreviations. Abbreviations are particularly troublesome because the target language may not allow abbreviations or translators may misinterpret the abbreviation. The icons shown in Fig. 6.16, for example, would require translation.

TABLE 6.6

Avoid Hand Gestures

Image	Intended meaning	Problems
	Yes, OK	In Sicily, this gesture invites the viewer to insert the thumb into a private part of the anatomy.
	Precisely, yes	This gesture can mean zero or worthless in France, and in South America it refers to a part of the anatomy not normally exposed in public.
	Stop, halt	In Greece, this palm-out gesture is considered obscene (Berlitz, 1982). The gesture dates back to Byzantine times when free men would humiliate new war prisoners by scooping up horse droppings and, using this gesture, rub them into the faces of the captives (Morris, 1977).

TABLE 6.7

Avoid Body Parts

Image	Intended meaning	Problems
	Move ahead	The underside of the foot can offend in the Orient. Showing bare feet can also seem crude and unprofessional or an accusation of poor hygiene.
	Inspect, examine	The single eye, viewed straight on, has many symbolic meanings. It is used on U.S. Currency to represent the all-seeing eye of God. Others may interpret the single eye as something evil, casting a curse on someone.
	Welcome	Take care with body language, especially conventional poses. This figure could be waiting to ask a question, hailing a taxi, delivering an insult, or giving a Nazi salute.

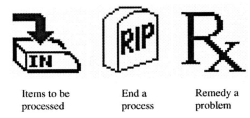

| Items to be processed | End a process | Remedy a problem |

FIG. 6.16.   Avoid words and abbreviations.

If you must include text in an image, style the text and image to make them familiar to users globally. For example, if you use a stop sign, shape the characters on the image exactly as they appear on the real sign. Use uppercase letters because they are more familiar to those whose language does not use these letters.

Initials and Individual Letters

Minimize using letters of the Roman alphabet as symbols. The emblems in Fig. 6.17 might work well throughout the Western Hemisphere and Western Europe where languages use the Roman alphabet, but could run into problems in countries whose languages use a different alphabet or symbol system.

Punctuation Marks

Likewise, avoid punctuation marks for symbols, as shown in Fig. 6.18. Many scripts lack these marks or assign them different meanings.

SECOND, LOCALIZE

Step 2 in our design strategy is localization. In localization we adapt the globalized version to make it suit the needs of specific languages, cultures,

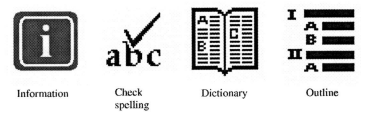

| Information | Check spelling | Dictionary | Outline |

FIG. 6.17.   Avoid initials and individual letters.

FIG. 6.18.   Avoid punctuation marks.

and locales. There are too many possible locales to offer specific advice for each, so this chapter will confine itself to general checklists and recommendations as to how the localization process for graphics can be made quicker and more reliable.

Design With Localization in Mind

Localization does not wait until after globalization is complete. The success of localization largely depends on designing the globalized version so that it can be readily localized.

Budget to "Translate" Graphics

Expect to translate some graphics. It is not unusual to translate 3–5% of the graphics for culturally distinct groups. Trying to be 100% international may make you 100% unclear. The solution is simple and relatively inexpensive: Redraw a few of your graphics. Changing 5% of your graphics is a bother, but it is often a better solution than the alternative of refusing to change the graphics and thereby confusing or offending users. It is certainly less expensive than avoiding graphics altogether and having to translate 100% of the equivalent text into multiple languages.

Train Designers and Illustrators

Teach designers and illustrators about cultural differences and how these differences manifest themselves in graphics. That way, designers can avoid problems from the start. Why not begin by routing copies of this chapter to all designers?

Leave Room for Text Expansion

As a rule, text expands when translated from English to other languages. This means you must leave extra room for text within graphics that will be translated. Expansion can be severe, especially for short phrases. The German for *popup message* has been translated as *Nachrichtenüberlagerungsfenster* (Kumhry et al., 1994).

Here are some ways to ensure adequate space for translation:

- Do not overlap the graphic with text.
- Leave plenty of space between labels and within boxes containing labels. Translated labels may expand and longer lines may wrap, causing the text to spill over borders or overlap other parts of the graphic.
- Position all labels, callouts, and other text outside the graphic so that the translated text will fit and not require redrawing the graphic.
- Allow the interface window to resize to accommodate expanding text.
- Do not use the smallest legible type size in the original graphic. That way, other versions can shrink the text to make it fit.

Create Localizable Source Graphics

Create your original graphics so that localizers can easily modify them and produce local versions. Here are some tips:

- **Draw graphics in vector rather than bitmap format** and provide the vector version to localizers. Vector graphics are easier to edit, to retype text, and to rearrange to fit text.
- **Layer graphics.** Place major parts of a graphic on separate layers, especially text for each target language, as shown in Fig. 6.19.
- **Consider file size.** For Web-based interfaces and content, consider the file size of graphics. Large files may provide more of a problem for users in locales without high-speed network connections.
- **Use standard, readily available fonts** in graphics. Be sure to supply localizers with all the fonts you use.

Provide Localizers Guidance and Source Materials

The team preparing the base, or global, version of the product, can leverage their efforts to greatly reduce the difficulties that will be encountered when it is localized. They can provide localizers with the information and source materials needed to produce local versions of the graphics. Here's a list of source material to provision localizers:

- Editable source graphics that localizers can easily modify to generate new local versions.
- List of anticipated localization problems to guide localizers in focusing their efforts.
- Specifications for graphics: file format, size, color palette, fonts, and any other details needed to produce compatible graphics.

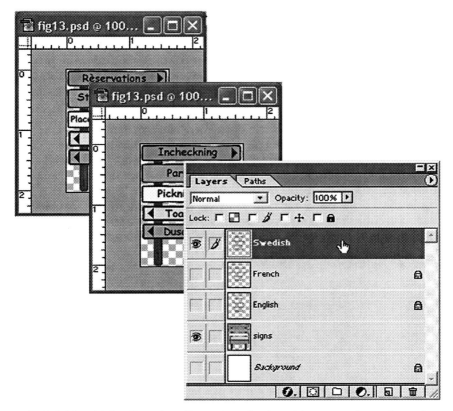

FIG. 6.19.   Adobe Photoshop illustration of a sign with separate layers for each target language.

- Lists of graphics that contain text or that you know will need to be modified.
- Alternative graphics for use where you know the original may not be appropriate. For instance, produce multiple versions of culturally specific graphics so that localizers have consistent versions from which to choose.
- Fonts to use in creating new text.

Adjust for Common Differences

Some aspects of graphics commonly require localization. This section provides some tips on identifying and handling these aspects of graphics.

Correct Provincialism

Some universal ideas lack universal symbols. The same general concept may be represented differently in different cultures. Consider the provincial symbols used to represent the universal concepts shown in Table 6.8.

Reformat Data

One area where we need to avoid provincialism is in data displays. For instance, your graphic includes a picture of a monitor showing a spreadsheet program. How are the figures formatted? Such displays of data need to be internationalized as well. Consider the kinds of displays shown in Table 6.9 as candidates for reformatting.

Formatting data may actually be worse because additional formats are widely used. Governments may decree one format as a standard; however, businesses may use another, and consumers may demand a third.

Adjust for Reading Direction

If you have images or icons showing bodies of text, these may need redrawing for languages that read in a different direction. Consider how each of the icons in Fig. 6.20 implies a left-to-right reading direction.

Remember, you will need to right-align labels when they are translated to Arabic or Hebrew, languages that read from right to left.

TABLE 6.8
Correct Provincialism

Mail		A rural route mailbox so common in the U.S. May not be a recognized symbol for mail (a widespread concept) elsewhere in the world.
Marriage		Although marriage is a common concept, what people wear vary from culture to culture.
Family		The notion of the nuclear family is a particularly Western concept. In many parts of the world, the notion of family may be more inclusive to encompass multiple generations and distant relatives.

TABLE 6.9

Types of Data That Need to Be Internationalized

Dates	Dates can differ in the names of months as well as in the order of the day, the month, and the year. 5/15/03 in the United States may be 15.5.03 in Germany or 15 mai 2003 in France.
Time	Time may be measured on a 12- or 24-hour clock and hours and minutes may be separated in different ways. The time 22 h 32 in France may be rendered as 10:32 PM in Canada and as 22.32 in other parts of Europe.
Currency	The units, currency symbol, decimal separator, and scale of values differ worldwide and may need to be changed.
Large numbers	Some countries use the period as a decimal point and the comma as a thousands separator. Others, however, reverse this convention. 123,456.78 in the U.S. Might appear as 123.456,78 in France.
Measurements	Even though the U.S. Uses the old British Imperial units, most of the rest of the world uses metric units. You may need to include both
Addresses	The placement of street numbers and postal (not ZIP) codes can vary considerably. In Germany, the street number follows the street name. In Japan, the city appears above the street address.
Phone numbers	The number of digits, how they are formatted, the presence of city and area codes, and special codes for toll-free calls vary from country to country.
Lists	Remember to re-sort alphabetized lists after translation. Don't forget that collating sequences vary (Jones et al., 1992). For example, in Swedish, the letter ä comes after z.

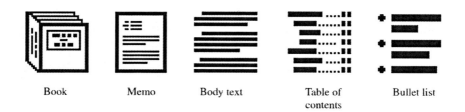

| Book | Memo | Body text | Table of contents | Bullet list |

FIG. 6.20.   Adjust text for reading direction.

Support Graphics With Translated Text

It is often less expensive to translate a bit of text than to redraw a graphic and much less expensive than to recover from a misinterpreted graphic. If graphics cannot be made universally clear, consider clarifying them with translated text. Such clarifying text may be:

- Labels within or next to the graphic (not ideal)
- Tool tips that pop up when the cursor points to the graphic
- Explanations in context-sensitive what's-this Help

Conduct Rapid, Informal Testing

Formal usability testing may come too late to help in localizing a product, and recruiting test subjects may be difficult and expensive if the target locale is distant. So, you may need to supplement your formal testing procedures with some of these inexpensive, informal procedures:

- Ask foreign sales offices test your interface over the Web.
- E-mail copies of images and their intended meanings to foreign sales offices. Ask whether anything in the graphics is inappropriate or confusing for users in their countries.
- Ask the distributors of your company's products throughout the world to review graphics for cultural issues.
- Hire representatives of target cultures. Seek out guest workers. Consulates and social clubs can often help you locate reviewers.
- Conduct a survey from your Web site. Offer premiums and discounts for helpful advice.

What If You Cannot Localize Everything?

For a complex product, designing graphics that are totally language independent and culturally neutral may prove impossible. Fortunately, users do not require perfect graphics, only recognizable ones. You can use images not common in the target culture provided your users can still recognize them. Most users would recognize the words *STOP* and *OK* on icons. Some objects, though not common, may be well known, especially among business people. For example, look at the objects in Table 6.10.

SUMMARY

Table 6.11 recaps many of the cultural and national differences you may encounter and what you can do to make your graphics more international.

TABLE 6.10
Use Objects Recognizable by Business People

 An Arab businessman will recognize this image of a book, even though books in Arabic, which reads right to left, would have the spine on the right.

 Likewise, his Japanese counterpart will recognize this image as a symbol of greeting among business associates, even though in the orient a bow is more common as a greeting than a handshake.

 A Chinese user will recognize this as an emblem for food even though chopsticks and a bowl may be more common in China.

TABLE 6.11
Recap of Cultural and National Differences

Difference	What to do
Racial characteristics	Use simple, abstract figures, devoid of recognizable bone structure or hair style. Use unshaded line drawings of people. Omit any indication of skin color.
Relations between sexes	Use simple unisex cartoons or stick-figure drawings of people, hands, and faces.
Clothing	Simplify drawings of clothing to omit seams, folds, buttons, and belts.
Modesty	Do not show bare arms, shoulders, legs, or feet
Gestures	Avoid hand gestures. If you show a hand, show the right hand holding or pressing something.
Color associations	Define and explain color symbology. Use colors in a technical context only.
Familiarity with graphical formats	Limit yourself to common, well-established formats.
Sense of humor	A whimsical style is OK, but avoid reliance on humor, especially puns.

Many of the issues fretted about in this chapter may mean less and less as time goes by. The World Wide Web, international commerce and travel, and global news organizations daily teach us more about each other's cultures, artifacts, and symbols.

REFERENCES

Axtel, R. E. (1991). *Gestures: The do's and taboos of body language around the world*. New York: Wiley.

Berlitz, C. (1982). *Native tongues*. New York: Grosset & Dunlap.

Boiarsky, C. (1992, September). *Strategies for successful international communication: Using cultural conventions to inform your documents*. Paper presented at the Record of the International Professional Communications Conference, New York.

Center for Universal Design. (1997). *What is universal design*. Retrieved May 20, 2002, from http://www.design.ncsu.edu/cud/univ_design/ud.htm

Cooper, J. C. (1978). *An illustrated encyclopedia of traditional symbols*. London: Thames & Hudson.

Deregowski, J. B. (1974). Pictorial perception and culture. In *Image, object, and illusion* (pp. 79–85). San Francisco: Freeman.

Dreyfuss, H. (1984). *Symbol sourcebook: An authoritative guide to international graphic symbols*. New York: Van Nostrand Reinhold.

Fernandes, T. (1996). *Culture4 and design in international user interfaces*. New York: Wiley.

Forslund, C. J. (1996). Analyzing pictorial messages across cultures. In D. C. Andrews (Ed.), *International dimensions of technical communication* (pp. 45–58). Arlington, VA: Society for Technical Communication.

Frye, R. H. (1981). Artistic technical training. *IEEE Transactions on Professional Communication, PC-42*(2), 86–89.

Fuglesang, A. (1982). *About understanding: Ideas and observations on cross-cultural communication*. Uppsala, Sweden: Dag Hammarskjold Foundation.

Gombrich, E. H. (1969). *Art and illusion: A study in the psychology of pictorial representation*. Princeton, NJ: Princeton University Press.

Grove, L. K. (1989). *Signs of the times: Graphics for international audiences*. Proceedings of the International Professional Communications Conference, New York.

Hagen, M., & Jones, R. (1978). Cultural effects on pictorial perception: How many words is one picture really worth? In R. D. Walk & H. L. Pick (Eds.), *Perception and experience* (pp. 171–212). New York: Plenum.

Halio, M. P. (1992, October). *Helping users navigate in multimedia documents: The affective domain*. Proceedings of SIGDOC'92, New York.

Hall, E. T. (1959). *The silent language*. Garden City, NJ: Anchor Press.

Hao, R. (1999). Really—How do Asians learn? *Performance Improvement, 38*(3), 13–15.

Hartshorn, R. W. (1989). *Writing for International Markets*. Paper presented at the Proceedings of the 37th Technical Writers Institute, Troy, NY.

Jones, S., Kennelly, C., Mueller, C., Sweezy, M., Thomas, B., & Velez, L. (1992). *Developing International User Information*. Bedford, MA: Digital Press.

Kano, N. (1995). *Developing international software for Windows 95 and Windows NT.* Redmond, WA: Microsoft Press.

Kittendorf, D. K. (1981, May). *The cartoon as technical communication.* Proceedings of the 28th International Technical Communication Conference, Arlington, VA.

Kostelnick, C. (1995). Cultural adaptation and information design: Two contrasting views. *IEEE Transactions on Professional Communication, 38*(4), 182–196.

Kumhyr, D., Merrill, C., Spalink, K., & Taylor, C. (1994). *International report: Concepts and procedures.* Durham, NC: Internationalization and Translation Services.

Merrill, C. K., & Shanoski, M. (1992, October). *Internationalizing online information.* Paper presented at the Proceedings of SIGDOC'92, New York.

Modley, R. (1937). *How to use pictorial statistics.* New York: Harper.

Morris, D. (1977). *Manwatching: A field guide to human behavior.* New York: Harry N. Abrams.

Mrazek, D., & Baldacchini, C. (1997). Avoiding cultural false positives. *Interactions, July + August, 1997.*

Neurath, O. (1936). *International picture language: The first rules of isotype.* London: Kegan Paul.

Ogilvy, D. (1985). *Ogilvy on advertising.* New York: Vintage Books.

Rahmat, O. (1999, July). I ain't no gringo. *New Media,* 27–36.

Rochester, J. (1992, September). *Visual aids for a foreign audience.* Proceedings of the International Professional Communications Conference, New York.

Spragins, E. S. (1992, October). *Developing hypertext documents for an international audience.* Proceedings of SIGDOC'92, New York.

Thorell, L. G., & Smith, W. J. (1990). *Using computer color effectively: An illustrated reference.* Englewood Cliffs, NJ: Prentice Hall.

White, J. V. (1990). *Color for the electronic age.* New York: Watson-Guptill.

Chapter 7

"Sunday in Shanghai, Monday in Madrid?!"

Key Issues and Decisions in Planning International User Studies

Susan M. Dray

David A. Siegel

Dray & Associates, Inc., Minneapolis, MN

INTRODUCTION

That you are reading this book suggests that you already understand why you might want or need to do international user studies. Suffice it to say that success in the international marketplace requires detailed knowledge about your international users and the context of usage, and evaluation of the design (preferably iterative) with representative international users. You must go beyond the generic information that is available about cultural differences. You must go beyond basic localization of your product, whether it is hardware, a Web site, or software. You must actually get empirical data specific to your users and your product. This is simply extending the principles of user-centered design to an international context.

Although international user and usability research is expensive, without this data to drive product definition and design, products have a much higher risk of failing in the international marketplace. Not only is the risk of a "miss" higher, but "misses" are more likely to be total. Over and over, we have found that international research has provided the type of "aha!" insights that made the difference between a product or system being viable or completely ill-adapted to the international context—insights that could not have been obtained any other way, except maybe by a product post mortem. Of course, it is much easier and cheaper to do it right the first time than to try to recover from mistakes by retrofitting changes to address international concerns or issues. So although this research is indeed more costly than domestic user research for many reasons, it is also indispensable.

If you are already an experienced international user researcher, we hope that it will be interesting and valuable for you to compare your experiences and approaches with ours, but we have a special sympathy for those new to international research. If you are looking forward to your first international usability evaluation or field research, you are likely to experience a mix of excitement and anxiety as you begin to think about the unfamiliar territory you are about to venture into. You begin to realize the number of new things you will have to manage and the likelihood that you will have to make various adjustments in your usual procedures to make the study "work" in another country, in another language, in another culture. Even more anxiety-provoking, as you realize how high the stakes are, is the fear that you will overlook a major issue in logistics or in methodology. You definitely do not want to find this out when you arrive in County A, with 3 days to complete the research before your scheduled departure for Country B.

We can empathize with both the excitement and the concern, and that is why we are writing this chapter—to give you the benefit of what we have learned through doing a variety of user studies, including usability evaluations and ethnographic research, in many countries in North America, Europe, Asia, and South America. We have tried to write the chapter we wish we had had available to us when we started with international work many years ago, and although this chapter is not a cookbook, it will introduce you to the issues you need to address as you plan and prepare for studies in a culture other than your own. We have organized the chapter around the key issues in preparation in particular, because preparation is the key to international user studies, and to managing the trade-offs that arise. We also briefly cover some of the things you need to know about actually conducting the study in-country. Of course, we focus on those things that are different from doing studies in your own country. In addition, because cost is usually a factor in planning these studies, we pay particular attention to planning issues that have a significant impact on cost, both explaining what drives these costs but also offering suggestions about strategies to make the research as cost effective and valuable as possible.

A Few Caveats

This chapter is based on our own experiences in doing user studies internationally and is intended to provide practical assistance. Although there is a vast literature on internationalization and on key cultural differences between countries, we do not review that here. Instead, we focus on the specific issues, especially trade-offs, that people doing international user studies need to address in their practical planning. We recommend and trust that, as a person interested in learning more about the countries where you will be working, you will consult the literature for more information and backgroud specific to those countries. This should be a part of your preparation for international user research.

   Additionally, our experiences are shaped, for better or worse, by our own cultural conditioning. Although we try to be sensitive to any biases this may lead to, we are also aware that we may not always see these clearly. Therefore, if you are not from the United States, and you find something here that is discrepant from your experiences, we encourage you to contact us so we may learn from you.

"It's All About Preparation"

It is a truism that you can't do too much preparation for an international user study. Preparation is always important, but it is especially crucial for conducting a study in a country outside your own. It is challenging to make arrangements remotely, time for problem solving when you arrive is limited, and it is unlikely you will have the opportunity to postpone or re-do the research with a modified plan if something significant does not work. It takes longer than you think it will to prepare and carry out an international user study, certainly longer than when you are doing domestic studies. This is true whether you are planning a visit study, such as an ethnography, naturalistic observation, or contextual inquiry, or whether you are planning usability evaluations. And even with all of your preparation, there will STILL be unexpected things to deal with once you get in-country.

MAJOR PLANNING AND PREPARATION ISSUES

Selecting Where to Go

When you are planning a study in another country, one of the first questions is often "Where are we going to test?" or "Where are we going to visit?" Sometimes circumstances dictate where you will have the opportunity to do the research, but often our clients are looking for some rational basis for making the decision. The question translates into some sub-issues like how to balance depth of coverage in one area with representation of different major geographical regions, and which countries are the right ones to visit in a given region. It is good to remember that one city in a country is not necessarily "representative" of that country as a whole, or that one country does not represent a continent. However, because you can never afford to go everywhere, you need some guidelines about how to sample.

Of course, in making this decision for a particular study, you have to be clear about what is driving the need for the research. This will often determine both the geographical scope, as well as whether the entire scope has to be addressed in one major study, or if there can be a series of geographically focused projects. Are there preexisting concerns specific to one geographical area, or is the product aimed at one particular overseas market? Obviously, this can limit the number of destinations you need to consider. Often, however, the product is intended for a global market, and there is a general need to better understand the global diversity of users and their responses to the design. Often, development life cycles mean that you will have one major opportunity for international research for a particular development cycle. Sometimes Web sites, such as corporate web sites, have a long life, undergo frequent updating, and have different national versions, creating opportunities for multiple independent studies with a more focused geographical scope, which can be conducted over time in an ongoing program of research. However, even for Web sites that are undergoing continual modification, there are usually limited opportunities for input into a major redesign. So, in these cases, it is important to make the most of the opportunity for international research, by making sure that your geographical sample gives you the maximum benefit.

Sampling for variation. Perhaps the key principle to remember when deciding where to do your research is that when you can not sample exhaustively, it is always good practice to sample for variation. This means that you allocate your resources to sample the most diverse locations you can manage. Obviously, this means diversity in relation to dimensions of difference that you have reason to believe are likely to influence the usage and usability of your product.

An important nuance in relation to this principle of sampling for variation has to do with the issue of what you can and cannot generalize from a study. On one level, you will learn things specific to the success of the design in a given location, and there will of course be a degree of uncertainty about how much you can generalize from these findings to areas you have not specifically visited. But on another level, because it greatly expands the universe of variation compared to doing research in your home country alone, international research almost inevitably exposes new dimensions of difference among users that you would not have anticipated. You may not know where people you have not sampled fall on that dimension, but once you become aware of a new dimension of difference, it becomes part of your general conceptual schema. You can at least take steps to ensure that it is addressed in your product planning and design research for other geographical areas. This is a form of "generalization" that increases the sophistication of your design research and thoroughness of your product planning for all areas. To put it in a nutshell, you don't have to see *every* manifestation of difference—you just have to see enough differences so you begin to understand how broad the universe of differences is and what some of the dimensions might be.

As an example, many studies have taught us that there are differing national patterns in such things as use of the postal system for sending payments or receiving shipments of products of any type, whether it is common for packages to be left at the door, and so on, and that these differences seem to relate to differences in e-commerce-related behavior and attitudes. Once sensitized to issues like this, you can add them to the checklist of things you consider when designing a product for any geographical area.

Though the principle of sampling for variation may be the primary guiding principle, there are many other factors that should and will affect your decisions about where to go. Here are some examples, along with some of the considerations they may introduce into your planning.

Cost. The cost of research varies greatly from region to region and country to country. In general, this is due not simply to differences in prices for services among countries, but also to factors that will influence how efficiently you can spend your time in the country. For example, in some countries, it is difficult or impossible to get working people into a research facility during working hours, so your research team will need to spend more time in-country for a given sample size.

Whereas actual costs depend on the actual research plan, and generalizations are possibly misleading, we can still offer some general impressions about relative costs. We have generally found costs in Asia to be higher than costs in Europe or North America, with Japan the highest cost area. The high cost of translation in Japan is a significant contributing factor. Costs in Singapore have tended to be an exception to the rule of higher costs in Asia, but this may be partly due to our circumstances as U.S. consultants, as the fact that it is often appropriate and feasible to work in English there greatly reduces our costs. We have found that costs in Latin America tend to be lower, although rental of computer or video equipment, if necessary, can be more expensive there than in the United States.

Obviously, you will have to take into account the differing costs of travel to different locations. Also, cost must be weighed against the relative value of going to particular places. Singapore may be less expensive than Seoul, but the same factors that make Singapore easier and less expensive may make it less interesting than a less "Westernized" city. Finally, another trade-off will be between the cost per location and the number of locations you can visit. You might be able to go to Mumbai (formerly Bombay) and Singapore for the cost of a similar study in Japan. In short, you will have to carefully weigh the trade-offs of cost versus the strategic importance of different destinations.

While considering costs, it is worthwhile to point out the cost efficiencies that can be derived from visiting more than one country per project. While the in-country research costs will be duplicated from one country to another, there are some "economies of scale." It is often possible to save on airfares by purchasing an around-the-world ticket, as opposed to separate round trips to several countries. Also, there will be some economies of scale in the planning effort for a project conducted as a single grand tour.

Market share. If you sell a lot of products in a particular country, this may be one appropriate way to focus your efforts. Going where you have significant market share will allow you to recruit people who use your product. Your product will also be more likely to be recognized or known even by those who don't actually own or use it, thanks to advertising and observed usage by others. Depending on this experience, as well as branding, and so forth, this may make it a good bet or a bad idea to go to a place where you have significant market share.

New markets. If you are intending to design a product for a new market, or you are introducing it into a new area, you may want to gather user information from that country or region. At the very least, it is important to gather ethnographic field data to help you in the earliest stages of product development to know whether your product even "fits" into that region, or whether you need to make fundamental changes in your approach. You will also want to check out your design decisions and do usability evaluations, preferably in an iterative manner throughout development, or, if you want to live dangerously, when you are approaching a final form.

Technology penetration. You may be able to decide which country or countries to visit based on the penetration of the technology in question, such as cell phones or the Internet. Depending on your circumstances, strategy might lead you to focus either on places where the technology is used a good deal, or places where the technology has limited penetration. Your market research department probably has a lot of information on this, especially if you are interested in more specialized technology penetration. For some things, you can find out for yourself on the Web, however. For instance, if you are looking for Internet use, you might want to check out http://cyberatlas.internet.com/. This site is a guide to online facts, gathered from a variety of sources.

Sales or field representative support. You may choose a country because you have a particularly helpful (or politically important) sales or field representative there. Sales/field representatives are often critical to recruiting, especially of current or potential customers and of enterprise-level customers. If you do not include them, you might find that the political support for your study can evaporate. However, when actively enlisted in your project, sales/field representatives can be incredibly helpful. They can help line up customers for field visits, identify cultural issues that will require localization of your test plan, help with finding equipment, and assist in translation by doing at least informal back-translations for you to check the accuracy of translated materials. They also know your company and its positioning within their country. It is important to educate them as to what you need and to help them understand what you are planning so they buy into your study. In some studies we've done, the client has chosen the locations based on where they had good sales/field support, and this was critical to the success of the studies. You may want to consider identifying ways to make it worthwhile for them to help you by including whatever perks and recognition are organizationally appropriate.

Sometimes the interest in the study comes from one area, such as when local field representatives have reported user problems in their territory. In cases like these, a particular field office may be the biggest advocate for user research. Using this as the entrée to international work may mean that you do not sample your entire international universe of users, but by working with these interested field representatives, you may build broader support for international research and set a precedent in your company.

Special challenges. If you know anything about conditions in a country that suggests your design will be challenged in a particular way, you should weigh that information heavily in prioritizing where to go, even if there isn't high market share or if it is more costly. For instance, if the product you are testing might work differently with different writing systems, you may choose to test it in a country with a pictographic system as well as one with a phonetic system. Or, as we mentioned earlier, you may want to test your e-commerce site in countries where there are known issues in people's readiness to order online.

Window of opportunity. You may find that you have a chance to piggyback an international study onto a trip planned for another reason. For instance, you may have to attend a sales meeting or go to a conference in another country. If you have the opportunity, you may be able to extend your trip to include a user study as well, thereby minimizing travel time and cost. Alternatively, you may be able to combine several studies in the same trip. For instance, we have combined usability testing of a product during the day with family ethnographies at night in order to collect as much data as possible in a short time (Dray & Mrazek, 1996). This is extremely exhausting, but it can help save money. Another possible synergy that can increase opportunities for international research while possibly reducing the effective cost, or even generating additional financial support, is to partner with marketing and coordinate international user-centered design research with marketing research. For example, as part of the same overall logistical effort, you might combine a usability evaluation study with a focus group study conducted on the same trip in the same facility.

Arranging for Research Resources In-Country

Once you have identified where you are heading, you must find and contract with the resources necessary in that locale to assist you. The solution you choose will probably be heavily dependent on whether your research will be done in a facility or in the field (homes or workplaces). Some of the services you may need to obtain in-country include facility rental, recruiting, translation (both of materials beforehand and simultaneous translation during the research), audio–video and computer equipment, and facilitation of the actual interaction with users. Unfortunately, working out this part of the plan and making your arrangements may be significantly more difficult than when you are doing something similar in your own country. Even though the Web has made it much easier to locate facilities and recruiting companies, the arduous task of vetting them and figuring out exactly who to use can still be quite time-consuming.

There are several different strategies or combinations of strategies for meeting these needs, and the appropriate solution for you will depend on many factors, such as the nature of the research, what resources your company may already have in-country to assist you, what your own experience has been with international user research, and what your capabilities are for conducting the research in a language other than your own. Here we focus on selecting and contracting with a facility. We focus on a scenario involving use of a market research facility, because this involves obtaining the most comprehensive set of services overseas and because it is the most common solution. Although you may be able to find a dedicated usability lab, commercial labs are still rare in many countries, and we know of no usability firms that can provide installed usability labs in a wide range of locations around the world.

Assuming you contract with a market research facility, you can at least expect them to be your single point of communication for arranging most or all other necessary resources. There are market research firms in most countries you might want to visit which offer one-way mirror observation rooms with fixed wall-mounted cameras and can provide or arrange for recruiting services, facilitation, translation, and interpretation. However, they will typically need significant guidance and oversight to prepare for user-centered design research such as usability evaluation, as compared to interviews or focus groups, with which they are probably more familiar.

In order to get meaningful bids from facilities, you have to ensure that your requirements are well understood. Pay special attention to communicating aspects of usability research these firms may be unfamiliar with. While recognizing that written specifications are only the start of the process, you must begin by creating as detailed a specification for them as you can, covering the research plan and requirements. Be sure to specify such things as a minimum configuration for the computer (processor speed, RAM, disk space, etc.) and/or Internet access (dial-up, dedicated or high-speed line, and minimum speeds) and audio–video requirements. In many market research facilities, the standard video arrangements include only a fixed focal length camera. If you need a more elaborate setup, such as direct screen capture through a scan converter and a "picture-in-picture" recording with an inset of the user, and assuming you do not have a portable lab, the facility will probably have to subcontract this. Because this setup may well be unfamiliar to them, you must explain it very clearly, so that they can obtain an accurate bid. Different countries have different video standards (PAL, SECAM, NTSC), so it is important to specify in advance which output you require. If it is not possible to record in your preferred format, you will have to plan for conversion of the videotapes. Depending on where you are from, it may be less costly to have these services provided locally and then have the tapes shipped to you, but if it is not feasible to have this done before you leave, and you are uncomfortable leaving the tapes behind, you will have to arrange for this back home.

It is essential to do everything possible to avoid misunderstandings about what a facility is bidding on, and given the challenges to communication that are inherent to international work, you must recognize the real risk of this occurring. As consultants, we have rarely found the level of detail we get from clients, even in formal RFPs (requests for proposal) to be sufficient for the purpose of obtaining bids from overseas facilities. This can be much less of a concern when we are working with a facility we already have a relationship with from previous work. However, we commonly find that we need to root out vagueness and ambiguity in these documents, usually by repeated conversations with the client. We then rewrite specifications in as much detail as we can, using simple, clear, straightforward English, and follow up with telephone conversations to ensure that the specifications and the nature of the research we want to do is understood.

Be prepared to find out that some of these requirements are not feasible as initially stated, or need to be adapted to the circumstances of the region. For instance, we have done testing of the same Web site using high-speed lines in some countries, and dial-up lines in others, because the local users in the latter would not have high-speed access to the Internet. Often, software will not run on older computers. Be sure to address this clearly in advance so you are sure that you will be able to conduct your tests once you arrive. As a final example, you may find that screener requirements are based on assumptions about the user demographics that need to be modified for the country (more on recruiting later).

If it is not already obvious, this entire process is extremely communication intensive. This is especially true if you are getting multiple bids from multiple countries. Allow the time needed for ensuring that facilities understand the requirements and for clarifying their responses. E-mail is helpful, but personal contact over the telephone is indispensable at key points in the process of negotiation, especially for getting a sense of how easy the interaction with the facility is. Depending on where you are trying to do the research, you will have to be prepared for some telephone calls at very unusual hours. You will also probably spend a lot of time discussing trade-offs among facilities with your team or client to make a joint decision about final location/facilities.

When you are looking for a facility to help with a user visit study or a usability evaluation, here are some of the things you need to consider:

Previous experience with this site. If you or anyone in your organization (such as people from a local field office) has worked with this group previously, don't overlook their input. What was their experience? How similar was previous work to that being considered now? Even if no one in your own organization has used them, do you know anyone who has? Can they give you references?

Claimed experience. Have they done similar types of studies? The concept of usability, and even the term, is entering the language of market research around the world. However, often claims of experience with "usability" do not mean what a professional trained in usability evaluation would expect (Siegel & Dray, 2001), and are more likely to refer to survey or interview research, where they asked about perceptions of usability, rather to behavioral research. Therefore, be sure to ask them to describe their usability research in as much detail as they can (not so much who they did work for, but what methods they used) so you can assess whether their perspective and definition of *usability* matches yours. Alternatively, it may simply be safer to assume that you will have to supervise the research, or arrange for a usability consultant to do so.[1]

One factor that may be a more relevant dimension of experience to assess in qualifying a facility is their level of experience in working with overseas clients. This is important for them to be able to understand your general level of expectations and the degree of coordination that will be necessary, as well as to give evidence of a track record meeting the needs of international clients, such as by providing bilingual facilitators and simulta-

---

[1]The same is true, by the way, if you are arranging for user visit studies. True user-centered design field research tends to be somewhat different from typical "day-in-the-life" or "ethnographic" studies conducted for marketing groups in companies. Therefore, it is important to check out what methodology the group has used in as much detail as possible.

neous translation. If this is not already an important part of their experience, look elsewhere.

A second experience factor to assess is their background in doing research on technical products and software, as opposed to household consumer products like foods and cosmetics. This provides at least some indirect evidence of their ability to arrange for resources appropriate to a usability study focusing on technical matters, such as facilitators comfortable exploring technical issues, translators who are comfortable with computer terminology, recruiters who can understand the point of your screener requirements, and so forth.

Location. Access to public transport is critical in many locales. How easy will it be for participants to get to the facility? Is it safe at all hours of the day? Are women and men equally safe? This can influence how successful the recruiting will be and may influence the schedule itself. Is it located near a reasonable hotel, and will it be easy for you to get to and from the facility? This is particularly important in usability evaluations because for visit studies, you are typically traveling to the users' locations, be they homes or businesses.

Your perceptions. When you interview a candidate facility on the phone, you will get a sense of their fluency and understanding. It is important to ask who, exactly, you will be working with because it may well not be the same person whom you are negotiating with. Interview the people with whom you would actually be working and see how easily you communicate. Look for the facility to ask penetrating questions about your requirements and to identify things that may have to be adapted for the study to work in their context. This is both part of the value that they add, and a way to check that they really do understand what you are trying to accomplish. A facility that confidently says, "Sure, we can do this" without raising questions or even objecting to some parts of our initial plan does not inspire our confidence.

Price. When you look over the bids you get back after this careful process, check to see if they are really comparable. Often, bids will include slightly different constellations of services and charges, so it can be difficult to make comparisons. If you have one "outlier"—either much higher or much lower than the others—check with that facility to make sure you have clearly communicated the requirements. Price alone is not a good basis for your ultimate decision, but it clearly does play a role. Don't automatically dismiss the high bid—they may be taking account of something that you will indeed have to deal with but that the others did not recognize or identify for you.

As we have already explained, testing in some countries, for instance, in Japan, can be much more costly than in other locales. If you need to test in high-cost areas, prepare yourself or your client for "sticker shock" and discuss with the facilities what things can realistically be cut back without compromising the actual testing. Some facilities, often smaller ones, may be willing to negotiate with you to find a mutually acceptable price.

Remember that the facility will be obtaining bids for some services, such as video equipment for screen-in-screen video. We have already emphasized the importance of making sure they understand your requirements. The other side of that is to make sure they are not bidding on more than you require. For example, we have questioned what seemed like high bids for video equipment, to find out that the bid included charges for three video staff (technician, producer, and director) to be on duty throughout the study. The price came down greatly when we made it clear those services were not needed.

Modifying a Recruiting Strategy and Screener to Reflect Local Cultural Issues

It is typical that you will find out while choosing the facility that some aspects of your initial plan don't make sense in one or more of the locales you plan to visit. This is the first step in localizing the test or visit plan. As mentioned previously, you want to be specific about your initial requirements, but be flexible to accommodate what you find to be the norm in that locale. You will want to ask the facility for their input on the feasibility of your initial goals for recruiting.

Some of the things you want to address include the following:

Adaptation of your screener. This is one of the most crucial tasks in preparation, and one in which there are interesting subtleties. The facility's recruiting staff can advise you on how to create a screener, or modify the one you have used in your own country, to find the people you need for the study. You may need to revise the quotas or requirements to reflect the actual demographics or usage patterns in a particular locale. Alternatively, your operational definitions, for instance, of "expert computer users" or "frequent" Internet users may need to be changed. Or the number of years that a person has been online may vary depending on the country, because the types of activities or applications used by "experts" may vary, or because the Internet is far less commonly used in some places.

In one example, we were working with a screener that required representation of different defined levels of Internet experience across socioeconomic levels. The facility provided statistics to show us that in their country, the distribution of home telephone service, and thus of dial-up Internet access, was such that the sample we described would have to be refocused toward the higher income levels. Without this information, the client would have been concerned about a sample that appeared to be skewed toward higher income levels than they had expected. Similarly, the difficulty of achieving balanced gender representation in different occupational categories can vary greatly from one country to another.

When you do research in more than one country, it is almost inevitable that you will want to compare findings across countries. However, there is no simple way to ensure that you are really sampling the same people in different countries. Rigidly insisting on a literal translation of the screener that you use for studies in your own country will not ensure "equivalence" because the meaning of that sample will differ across countries. When the local context seems to require changes in how you structure your sample, you will have to carefully consider how to be sure you are sampling the group you are really interested in. You will need to think about whether your screener concepts are absolute objective definitions, or if they are culturally relative definitions. Be prepared to take some time for conceptual negotiation with your client or team, discussing whether adjustments in the screener are giving you a sample that is the analog in this context of the sample you might put together somewhere else. Also, remember that this process may give you insights about real differences in who your users are in different countries, insights that you might want to follow up on with quantitative data.

It is obviously most efficient for the recruiter to assist you in modifying the specific requirements before the screener is translated so you have an opportunity to reword the screener. When the screener is translated, the recruiter may reorganize it into their standard format. Review it very carefully to make sure that the logic of the decision rules remains intact and that the key concepts have not been distorted in translation or reorganization. This is not a place to skimp on back-translation, which we discuss later.

Recruiting strategy. The recruiter can also tell you the best way to find your target users. In some countries, cold calling by recruiters is illegal or not effective. Recruiters in these countries know how to get people to come in for testing, however, as well as how many they need to recruit for a given quota of completed tests or visits. They will also know how much you need to pay or how else to provide effective incentives to increase your show-rate or visit acceptances.

Just as some recruits are more difficult in your own country, the same is true in other countries. The facility should be able to advise you on the ways to reduce the cost of challenging recruits. Sometimes, if you can compromise your recruiting strategy somewhat, either reducing the total number, or relaxing restrictive requirements, you can make the recruiting easier, and therefore, less costly.

In some countries, partly due to the difficulties of getting working people into facilities during weekdays, there is a tendency for samples to be overly loaded with students and housewives. There can also be a very blurry distinction between people who are self-employed and people who are unemployed. This is something that we address explicitly, by setting quotas and by negotiating the definitions of employment status criteria.

If you are doing a visit study, discuss with the recruiter whether your plans are realistic. Will you be able to get people to allow you to come to their homes or offices? If not, are there alternatives, such as doing an interview with them on the phone or at the facility initially to build rapport, and then visiting them? Can you do without recordings, video or audio, if that is a stumbling block? Do you need to make changes in the length of the interview, or the size or composition of the visit team to accommodate cultural issues? We typically find that a cross-cultural visit team is most effective, but you need to make sure that this will be acceptable in a particular locale. Also make sure whether you can have a mixed gender team. In some places, this may be less acceptable. Because the purpose of visit studies is usually to understand the context of the users, it is critical that you identify ways to get this information, even if it means modifying your initial plans.

Scheduling. This goes hand-in-hand with recruiting. It is understandable that your team may want to spend as little time in-country as possible to minimize time and travel costs. However, in many locales, if you want working people as evaluators, you must run evaluations in the evenings. In some places, you can also run on weekends, but in others this is never done. For working people in Japan, for instance, you must plan to schedule on the weekend because it is difficult to schedule during the week. In contrast, in some Latin American countries, we have found, we were able to do usability evaluations weekdays and evenings, but not on the weekends. Your facility can help you identity a workable scheduling plan. Your specific user study protocol will also influence scheduling. Often, it is possible to combine usability testing and ethnographic visits by using different time slots. For example, in a project in South America, where it was difficult to do facility-based usability evaluations on the weekend, it was fairly easy to do home visits on the weekend.

If you are visiting families, you need to think about when all needed members of the family will be present (assuming this is important to your study), and what you can do to "break the ice" and build rapport as quickly as possible so you can get the information you need. When we do family studies, we often take food in with us so we can meet earlier, over the dinner hour, for instance. This is a common practice in some places, such as the United States, but is very unusual in other locales. It can also be difficult to make arrangements for this in some places. For instance, when we did family visits in Germany, we had difficulty finding a restaurant that could provide food "to go" for us (Dray & Mrazek, 1996). Although this was an unusual practice, we found that it worked well, especially as it was the "crazy Americans" who were bringing the food. It might have been more difficult for a local person to break the norm in this way.

Planning for no-shows. Whenever you plan a usability evaluation, you have to plan for no-shows. Rates of no-shows vary widely from locale to locale. In some places, no-show rates are extremely low, so scheduling one extra slot, as a backup, is all that is necessary. Other places, however, have predictably high no-show rates. These may be influenced by specific local conditions, such as whether it rains (making transport slower), or what day it is. As another example, payday may be a bad day to schedule because in some places it is common for people to pay their bills in person that day, snarling traffic and making travel times unpredictable.

There are a variety of ways to deal with high no-show rates. Some facilities simply advise scheduling multiple extra time slots. Others use "floaters" as backups. In some places, facilities typically recruit multiple people per slot (as many as three) to deal with high no-show rates. This can make it more difficult to control the sample in terms of segmentation unless they are able to recruit multiple people from the same segment for the same time slot, which is unlikely and impractical to require. It adds to the difficulties in exporting your segmentation model, which we have already discussed. When multiple people show up for the same time slot, and you have an opportunity to choose, it is therefore very important that you carefully review their characteristics against your quotas and against the range of people on the list for the time slots yet to come, to make best use of the degrees of freedom that remain to you.

No-shows for visit studies are less common, but not unheard of. It is much harder to overschedule, given the amount of time each visit takes. Therefore, when we are doing visit studies, inside or outside our own country, we routinely call the day before to answer questions, check the instructions for reaching the user(s), and establish preliminary rapport. If visiting a company or office, we also send ahead any nondisclosure or informed consent form so that these are not a shock to participants. This helps to establish that we are, indeed, legitimate, and makes cancellations less likely.

Planning for extra time. When scheduling your first test or visit, it is important to allow the team some time to adjust to a different time zone. Usually, we try to travel so we have at least a day, and preferably a weekend, prior to the beginning of the study to adjust, although sometimes teams need more than a single day for adjusting to both time and culture if they have not been in this locale before. A good compromise is to schedule an equipment check and dry-run for the first day. We discuss these activities in more detail later.

In addition to initial time to adjust to new time zones, it is important to allow more time between sessions than you might expect to spend in your own country. The process of doing usability or ethnographic studies in a country other than your own, especially one with a language you don't speak, requires much more concentration than one in your own country, even if you are very experienced. You will find that it takes more time to process what you are observing, and this needs to be planned for. In addition to the things you normally do in a debriefing, it is important to have the time to compare your impressions with those of the facilitator and interpreter(s). For example, it may be necessary to review how the interpreter translated some key ideas to make sure you really have captured the sense of the words. This is also a valuable opportunity for coaching the facilitator or planning any modification for the next session.

Preparing or Adapting the Protocol and/or Test Plan

A critical step is to adjust the specific methodology to the specific locale, balancing carefully the need to keep things as similar as possible, for comparison purposes, with the need to modify to fit local culture. This needs to be done at multiple levels, from the overall goals to the specific tasks you plan to use.

Goals and underlying questions. Obviously, the first step in localizing a test plan is really understanding what the underlying questions are that must be answered and then figuring out how to get at those answers in the context of each culture. If you are the one who developed the test plan, this may seem obvious. However, if you did not develop it yourself, or are working with a team where there may be different perceptions of the critical information, a crucial first step is to become intimately familiar with the issues the team needs to understand, so that the focus will remain intact when you localize the test plan. This is always important in user studies, but especially so in preparing for international work. Because of all the other factors that may divert you, you will need an especially strong conceptual compass.

Tasks. You may need to modify task scenarios to reflect more realistic or typical tasks, especially if you are not basing your tasks on previous ethnographic visits. At the very least, certain elements, such as data to be entered, names, and/or address formats, will need to be localized. However, entire tasks may need to be adapted to the local context. It is beyond the scope of this chapter to discuss the many issues involved in how you select task scenarios for usability evaluations. For our purposes, it is sufficient to simply underline one key issue: the distinction between choosing tasks to represent typical user activities versus choosing tasks specifically designed to probe certain aspects of the design. Sometimes, these two different ways of choosing or prioritizing tasks do not overlap. A problem can arise when a task that you think of as a typical activity is actually unfamiliar to users, uncommon in their context, or presented in ways that do not match how they typically think about the task. We generally want user difficulties to reflect design issues. We do not want them to be mere artifacts of problems with understanding the task instructions.

For all of these reasons, it pays to discuss the specific tasks you plan to use in usability evaluations with the facilitator (again, before they are translated if at all possible). However, it is important to have realistic expectations. Although the facilitator can and should help with localization of terminology, data formats, and so on, and although they may be able to raise concerns about some of the tasks at a conceptual level, do not expect them to be expert spokespeople for describing the technology usage patterns and ways of thinking of all their fellow citizens, especially if you are targeting a specialized population. For some applications it can help to have your tasks reviewed by a local person who knows the domain, if possible. Of course, we should never rely too heavily on a single informant. To some extent, you will have to be prepared to modify task scenarios based on your observations of how they are actually working during the study. Learning that you have to change your assumptions about task scenarios can in fact be one of the big benefits of your research.

Methodology. You may need to consider changes to your methodology to adapt it to cultural constraints. These include such things as differences in how people express their subjective experience, thoughts, and reactions in a social context. There are also cultural differences in how ready people are to express criticism or confusion, or to attribute problems to the design rather than to their own failings. In addition, how inhibited people may be in the presence of a foreigner seems to differ from country to country. All of these factors influence how well such basic techniques as "think out loud" will work. Concerns about these cultural factors are compounded by the need to adapt your methodology to working through the intermediary of a facilitator who is often not a usability professional. All of these considerations have a tendency to lead people to think that only a more structured, less interactive, more quantitative or "metric" evaluation approach is feasible. This is unfortunate because the team may need the kind of design guidance that is more likely to come from a more interactive exploratory or "formative" approach to evaluation than from metric testing, which focuses on assessing success in task completion and may be more appropriate as a "validation" test. Furthermore, if you are planning ethnographic work in other countries, you have no choice but to find a way to facilitate rich interactive exploration with users, despite cultural and language barriers.

A thorough discussion of these issues and how to deal with them is beyond the scope of this chapter. However, if you are interested in highly qualitative forms of user and usability research, we urge you not to give up on these too readily. Fortunately, we have found that, with careful planning, choice of method, and preparation of the facilitator, it is possible to use flexible, interactive approaches, such as various exploratory usability evaluation methods and, in the realm of ethnography, contextual inquiry. For example, in a country like Japan, commonly considered a context where "think-out-loud" does not work very well, we have found that co-discovery can be effective at eliciting verbalizations that give the needed clues to the user's cognitive processes. Your facilitator should be able to help you plan alternative strategies (once the nature of usability evaluation is understood). For example, when we first used co-discovery in Japan, our facilitator helped us realize that it was important that the two participants be of equal social rank, or the one who was subordinate would not speak freely. We had the Japanese facilitator assess social rank based on cues that to us, as non-Japanese, were difficult to assess, but to her, were obvious.

Working through a local facilitator is often part of the solution to cultural barriers, but at the same time, it introduces methodological issues of its own. If the test is a highly structured metric test, then real-time interaction with the user will be less, and the intermediary role of the facilitator will be more structured or more narrowly "procedural," and thus in a sense less influential and easier to manage. This contributes to the tendency for some people to believe that international evaluations almost necessitate this type of approach, in contrast to exploratory evaluation methods. For example, it may be assumed that the facilitator's role should be limited to introducing the test procedure, handing the user cards with written tasks, and then intervening minimally and only if the user becomes "stuck." In contrast, we have found that it can indeed be possible to engage in a much richer qualitative evaluation, even when working through a facilitator. This depends on thorough preparation of the facilitator and on choosing a workable method for communication with the facilitator during the data collection. We discuss these issues later in the chapter.

Arranging for Translation

We have already briefly referred several times to the importance of translation. In this section, the issue of bridging between languages is our primary focus. Bridging between languages involves three components: facilitation, translation of written materials, and simultaneous translation (also called *interpretation*) of the sessions themselves.

Bilingual facilitation. We say more about the role of the facilitator shortly, but in relation to the language issue, here we simply want to point out that it is crucial for the facilitator to be able communicate both with the team and with the users. The facilitator has to be comfortable enough in both languages to communicate and understand nuances, or you will have to communicate through a translator.

We believe that all testing should be done in the local language. Only rarely, in our experience, has it been defensible to use English when working with people for whom English is not their first language, such as when working in a technical domain where English is in fact the language of everyday activities for users. However, even with people like this, not all of the verbalization you want to capture consists of technical content. You want people to express their thoughts and reactions as spontaneously and freely as possible, which argues for providing facilitation in the local language.

A compromise solution sometimes attempted is to facilitate the evaluation in your own language, assisted by an interpreter when necessary. If you intend to use the interpreter for backup only, this may require you to skew your sample by requiring that participants speak your language. The idea of facilitating in your own language, with all communication mediated by the interpreter is also attempted sometimes, but can be even more problematic. Not only does it greatly impair the immediacy and spontaneity of communication, but it is laborious, makes it difficult for any probes to come at the right moment, and creates distractions for the user. Furthermore, as we have already mentioned, in some countries, users are much more likely to interact freely and openly with a member of their own culture.

These problems bring us back to the desirability of providing facilitation in the local language. This creates the need for simultaneous translation, which we discuss next. Of course, it also presents challenges in regard to selection and preparation of the facilitator, and we discuss this area in a subsequent section.

Written translation. All of your materials, starting with the screener and including all documents for the testing or visit itself, will need to be translated into the local language. This is typically a job for a professional, preferably one who is familiar with technical terms that may be included in your materials. The translator may be associated with the facility you use, or may be an independent or freelancer. In either case, your logistics plan has to allow time for this step. It is difficult to get a reliable bid on this type of translation because it is charged on a per page basis, and at the time you are getting bids, you may have only a vague idea of how many pages will be involved. In our experience, however, this is not an especially costly item.

One document that is part of most protocols is an Informed Consent/ Non-Disclosure Agreement (NDA) for the research participants to sign. Facilities typically supply their own "standard" versions of these forms, which might avoid the need for translation, but you should check to make sure their version meets your requirements. You may decide that you need to use your own form for your purposes, even if they automatically have participants sign their form when they arrive.

It is common to back-translate materials to make sure that they are accurately translated. This can be seen as an added cost, but is a good idea, especially if you do not have access to someone who is fluent in the target language(s) who can give you an informal check to make sure the translation has captured the spirit and nuances of the original.

Often, the project timeline requires that work begin on translation of materials while the study methodology is still evolving. This can result in multiple cycles to update the translations. This makes version control extremely important. It is amazing how rapidly different versions can multiply and potentially cause confusion, especially when multiple people in different countries all need to be looking at the same version at the same time. The challenge is compounded when you are preparing a multicountry study.

Simultaneous translation of sessions. The simultaneous translator, or interpreter, provides a voice-over that observers hear through earphones and which is also recorded as a track on the audio- and videotape of the sessions. The interpreter may be in a soundproof booth in the observation room, may be sitting behind the team in the observation room, or may be monitoring the session remotely via a video hookup or even audio only. In any case, this person must be highly fluent and able to maintain concentration for long periods of time, passing along every comment, and conveying nuances in real time to the team. Skill in translating documents is not the same as skill in simultaneous translation. Clearly, this is a job for a highly skilled professional. As a result, this critical role is also often quite costly. The costs vary from market to market, with Japan being the highest in our experience. Translation costs can often be equal to or greater than the daily facility rental costs. However, do not be tempted to cut corners on the interpreter, as problems with interpretation can undermine the value of the entire project.

Cost is also increased by the fact that, in some countries, interpretation is commonly done by a team of translators, who can take turns during a lengthy session, or who can alternate sessions during a long day. Even when this is not the case, it is critical to allow enough time between sessions to give the interpreter a break. Also, plan food and refreshments for this person or team during the breaks, so they can maintain their focus on the translation.

Training Local Facilitators and Preparing Facilities

A key to success is to make sure that the local personnel are fully trained in the methods you plan to use. This is especially true if you are using a local facilitator who is experienced in market research methods, but not in usability. Whereas part of this takes place before you arrive, part of it can only be done after you are on the scene.

Training of local facilitators in usability methods. Assuming that you are not working with a trained usability professional who is fluent in the local language, you will have additional work to prepare the facilitator. As we have already mentioned, the role of and demands on the facilitator depend on the style of testing you plan to do. If the test is a highly structured metric test, then you may be able to focus primarily on training the facilitator in the mechanics of the evaluation process. However, if you plan to conduct a more exploratory usability evaluation, you will need an appropriate process to train the facilitator, as well as a strategy for communicating with the facilitator during the session.

When we are working with a facilitator for the first time, we typically provide training in the concepts of usability evaluation, both before arrival, and via hands-on training, role playing, and dry runs on site. We usually spend time on the phone with the facilitator prior to our arrival, giving him or her an overview of the concepts, with a focus on how usability evaluation differs from methodology they may be more familiar with. We also send written materials and pointers to Web sites, including materials we have available on our own Web site, to give them a background in what will be expected. Depending on the circumstances, we have used some other strategies. Sometimes it is possible to send a video of a related test so they can see how we have done it. This works best if it is a tape of an earlier evaluation of the same product. Occasionally, we conduct a mock session with the facilitator over the phone, using a public Web site and a set of standard tasks. For instance, we may have them think aloud while trying to find and order a book at a popular bookselling Web site in their country, or have them find some information, such as the weather forecast, for their hometown.

Once we arrive, we go over this information again and then conduct a dry run in person. During the dry run, we often will first demonstrate the process with the facilitator playing the role of the user, then have them conduct a session, with one of us taking the role of the evaluator. If possible, and appropriate to the technology being evaluated, we end by having the facilitator conduct a dry run using a facility staff member as the user. This process is valuable on several levels. Of course, it helps the facilitator learn the test plan thoroughly, while at the same time revealing last-minute modifications that may be necessary. Second, it is an opportunity for the facilitator to learn the paradigm of usability testing with its focus on user behavior and cognitive process rather than opinion, and some of the basic methods of exploratory usability evaluation, such as ways to elicit thinking out loud and ways to probe nondirectively. Finally, it is also a chance for them to understand the underlying goals and questions that are implicit in the test plan and get an idea of the types of issues that interest us in relation to this particular design. This is critical, as it helps them get a sense of where to take the initiative to follow up and draw people out to help the team answer key questions.

This level of preparation has proven its worth many times. It provides facilitators with a good sense of what to expect before getting into the actual usability evaluation sessions, and helps them function much more effectively as collaborators. It also helps make the entire process much more rewarding and professionally stimulating for them. Furthermore, there may be a longer range benefit from the effort you put in with your facilitators. In our case, people we have trained for this role have become part of our network of collaborators, allowing us to do subsequent international studies much more easily. The same may apply for you.

The preparation prior to the research, however, is not enough to turn the facilitator into a usability expert. In addition you need to plan for how you will communicate with the facilitator during the session when necessary, or even if you will attempt to supervise and direct the process of testing from behind the mirror. Some of the methods that people use are passing notes, breaking for a quick consultation during the session, or even knocking on the door and calling the facilitator out. These methods work best in situations where there are clear predetermined break points, or there is a fairly clear threshold for intervention, such as when the process has clearly broken down and the facilitator needs some guidance about how to get back on track. However, they lack the flexibility and immediacy necessary for probing those ephemeral opportunities for judicious probing that tend to occur in usability evaluation and often lead to deep insights—moments which the facilitator may miss.

To address this, we have often used a technique called the "bug in the ear." This is a wireless intercom that uses an earplug speaker/microphone. It allows the usability expert behind the mirror to communicate immediately to the facilitator. However, this takes practice. If this is not used skillfully and judiciously, in terms of timing, quantity, and wording of messages, it can be potentially confusing and disruptive to the facilitator. However, we have found that, given some time for familiarization and practice, it can be very effective. We routinely use this technique when the two of us work together on a usability evaluation, with one serving as facilitator and one behind the mirror and, therefore, have had the opportunity to practice ways to time our communications and to word specific probes to be most useful to the facilitator.

If a facilitator is not comfortable with the "bug in the ear," if the person's level of comfort with your language is not high enough, or if the technique is not working effectively, we have resorted to other compromise solutions, such as the common ones of passing notes under the door or having the facilitator step out of the evaluation room for clarification. Depending on the risks of our being present with the user, we have also sat with the facilitator, to pass notes when necessary. Sometimes we have done this while wearing the bug ourselves, so we can relay suggestions from the observers (subject to our own "editorial" authority). Finally, even if the bug in the ear can not be used in a rich and flexible way to cue the facilitator for specific probes, it can still be quite helpful when used in a very limited way to help manage the flow of the session, such as by using it simply to signal the facilitator to move on, or reminding the facilitator to encourage thinking out loud. It is helpful to agree in advance to a small number of simple code words you will use for these purposes.

Checking and correcting problems in equipment set up prior to test. It is helpful if you can be present when the equipment is being set up. For instance, if you have arranged for a picture-in-picture recording, it may require some tweaking to set up properly. Because you know what aspect of the screen is the least likely to have critical interactions, you may need to specify where to position the inset. It also helps to be present at setup so you can understand what to do should troubleshooting become necessary in the middle of a session, since, typically, it is not cost effective to have subcontracted video technicians stay onsite for the duration of testing.

Train interpreters. Another key to success is briefing interpreter(s) on the process of usability testing. This allows them to understand more fully why you need to hear everything, and why even offhand comments can be the keys to user confusion that you need to understand in order to make changes in your product. Because of the cognitive load on them, interpreters may normally let these chance comments go by without translating them, so it is especially important that they understand the value of these comments to you.

We usually conduct our training of interpreters shortly before the actual sessions. We give an overview of usability evaluation (or ethnographic visits, if that is what we are doing) and explain carefully the types of things that are most useful to hear. We try to help them understand that the tone of voice, as well as the words, is important to convey. This helps the interpreters to balance their load, by focusing on the relevant nuances of tone or expression that are most useful. We try to have the interpreters present for the dry runs that we do with the facilitators, and may even use the interpreter as an evaluator in a dry run to help them get a better sense for the setup. If the interpreter is observing the evaluations only through a video connection (or if they cannot see it at all) this helps them to visualize the set up, which can help with their interpretation.

Multicountry Studies: Serial or Parallel?

We have found that it is very beneficial to have continuity on the observation team for multicountry studies. This means that at least one key person needs to observe all of the studies—visits or usability evaluations—in all of the countries. This may seem more costly, but in reality, it is very helpful to ensure that you can make real sense of the findings. Even though it is valuable to have "new eyes" along as well, it is critical that someone has seen all of the users so that any differences in procedure or result can be put in context. In studies where this is not the case, it can be extremely difficult to know whether to attribute a difference in findings to a difference between locales or to an inconsistency between teams.

In addition, what you see depends on what is salient to you and this in turn is influenced by the patterns you have already seen. Things that would not have been noticed by themselves can become noteworthy when they stand out against other patterns. This is the familiar figure/ground phenomenon. For example, in a recent study in two countries in the same larger geographical region, we noted that in one country people tended to make an initial attempt to guess at a URL, whereas in the nearby country, when faced with the same web navigation task, almost all the users followed a strategy of surfing links starting from their home page. In both cases, this observation was incidental to the main point of the study. Therefore, if we had not seen these two contrasting patterns side by side, the chance that we really would have taken note of them would have been less.

It goes without saying that if you are to have continuity, it is not recommended that you conduct your study in different countries in parallel. Of course, it takes more time to do them serially. If time is of the essence, you may want to examine other ways to expedite the data gathering, before resorting to parallel teams, such as reducing the number of visits or evaluations, or selecting the locations to allow more rapid travel between them. If you can not provide the continuity yourself, consider having a consultant play this role.

However, if it is simply not possible or feasible to conduct the study in series and to maintain some continuity across teams, then you will need to devote extra effort to planning for some standardization of methodology, and ongoing coordination among teams. We recommend that you plan to maintain communication with and among the teams during the data gathering, so that modifications in the procedure can be coordinated as appropriate, and insights and possible trends can be shared. This can help reduce the risk that if two teams come back describing different themes and trends, it is not merely because their thinking happened to get into different channels and they were not aware of what the other teams were doing.

CONCLUSIONS

In this chapter, we tried to give you a flavor for some of the major decisions and trade-offs that you are likely to encounter when doing international usability evaluations or ethnographic visits. Certainly there are many other issues one could discuss, but it is not possible to cover them all in one chapter, and these are some of the biggest ones. It would have been easier for us had we known these things in advance, rather than having to learn them the hard way. The community of professionals doing international user research is growing, and we therefore have increasing opportunities to learn from each other. With this in mind, we welcome a dialog with you as you pursue your quest for international usability.

REFERENCES

Dray, S. M., & Mrazek, D. (1996). A day in the life: Studying context across cultures. In E. M. del Galdo & J. Nielsen (Eds.), *International user interfaces* (pp. 242–256). New York: Wiley.

Siegel, D. A., & Dray, S. M. (2001). New kid on the block: Marketing organizations and interaction design. *Interactions, 8*(2), 19–24.

Chapter 8

Cost-Justifying Usability Engineering for Cross-Cultural User Interface Design

Deborah J. Mayhew
Deborah J. Mayhew & Associates, West Tisbury, MA

Randolph G. Bias
University of Texas at Austin, Austin, TX

INTRODUCTION

In this chapter, we discuss the cost justification of adding usability engineering to projects aimed at developing software to be used in multiple different countries and cultures. Although both of us have done significant work in the area of cost-justifying usability engineering in general, neither of us has significant concrete experience with designing for cross-cultural audiences. We have drawn on the writings of and conversations with our colleagues who do have significant experience in cross-cultural design, and extrapolated from our collective experience about cost justifying the application of usability engineering in cross-cultural contexts.

Cost justification of usability engineering in general is addressed in detail elsewhere (Bias & Mayhew, 1994; Bias, Mayhew & Upmanyu, 2003; Mayhew & Bias, 2003). Here, we focus on what is unique about adding usability engineering to a global project, and how to adapt the general cost justification technique in this case. To do this, we first provide an overview of the general usability engineering process. Then, we provide some discus-

213

sion and examples of unique aspects of the usability engineering process for cross-cultural projects. Based on this background, we then provide an overview of the general cost justification technique and some examples of adapting the technique to cross-cultural projects.

Overview of a Generic Usability Engineering Lifecycle

The Usability Engineering Lifecycle (Mayhew, 1999) is a structured and systematic approach to addressing usability within the product development process. It consists of a set of usability engineering tasks applied in a particular order at specified points in an overall product development lifecycle.

Several types of tasks are included in The Usability Engineering Lifecycle, as follows:

- Structured usability requirements analysis tasks.
- An explicit usability goal setting task, *driven directly from requirements analysis data.*
- Tasks supporting a structured, top-down approach to user interface design that is *driven directly from usability goals and other requirements data.*
- Objective usability evaluation tasks for iterating design toward usability goals.

Figure 8.1 represents in summary, visual form, The Usability Engineering Lifecycle. The overall lifecycle is cast in three phases: Requirements Analysis, Design/Testing/Development, and Installation. Specific usability engineering tasks within each phase are presented in boxes, and arrows show the basic order in which tasks should be carried out. Much of the sequencing of tasks is iterative, and the specific places where iterations would most typically occur are illustrated by arrows returning to earlier steps in the lifecycle. Following the chart are brief descriptions of each lifecycle task.

Phase One: Requirements Analysis

In this phase, user research is conducted and collated. A description (User Profile) of the specific user characteristics relevant to user interface design (e.g., computer literacy, expected frequency of use, level of job experience) is obtained for the intended user population. This will drive tailored user interface design decisions and also identify major user categories for study in the Task Analysis task. A study of users' current tasks, workflow patterns, and conceptual frameworks is made (Task Analysis), resulting in a description of current tasks and workflow and an understanding and specification of underlying user goals. These will be used to set usability

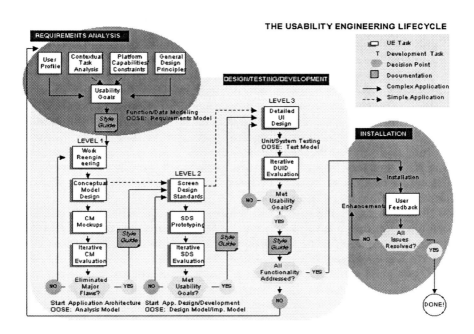

FIG. 8.1. The usability engineering lifecycle (Mayhew, 1999, used with permission).

goals and drive Work Reengineering (also known as Information Architecture) and user interface design.

The user interface capabilities and constraints (e.g., windowing, direct manipulation, color, etc.) inherent in the technology platform chosen for the product (e.g., Apple Macintosh, MS Windows, product-unique platforms) are determined and documented (Platform Capabilities/Constraints.) These will define the scope of possibilities for user interface design. Relevant general user interface design guidelines available in the usability engineering literature are gathered and reviewed (General Design Principles.) They will be applied during the design process to come, along with all other project-specific information gathered in other tasks in this phase.

Based on all usability requirements as described above, both *qualitative* goals reflecting general usability requirements and *quantitative* goals defining minimal acceptable user performance and satisfaction are developed (Usability Goals.) These usability goals focus later design efforts and form the basis for later iterative usability evaluation.

Phase Two: Design/Testing/Development

Level 1 Design. Based on all requirements analysis data and the usability goals extracted from them, user tasks are redesigned at the level of organization and workflow to streamline work and exploit the capabilities of automation (Work Reengineering). No user interface design per se is involved in this task, just abstract organization of functionality and workflow design. This task is sometimes referred to as Information Architecture. Next, initial high-level design alternatives are generated. At this level, navigational pathways and major displays are identified, and rules for the consistent presentation of navigational controls, work products, processes and actions are established (Conceptual Model Design). Screen design detail is *not* addressed at this design level.

Mockups (paper-and-pencil or live prototype) of Conceptual Model Design ideas are prepared (Conceptual Model Mockups.) Detailed screen design and complete functional design are *not* addressed at this stage. The mockups are evaluated and modified through iterative evaluation techniques such as formal usability testing, in which real, representative end users attempt to perform real, representative tasks with minimal training and intervention, imagining that the mockups are a real product user interface (Iterative Conceptual Model Evaluation.) The design, mockup, and evaluation tasks are conducted in iterative cycles until all major usability "bugs" are identified and engineered out of the Level 1 (i.e., Conceptual Model) design. Once a Conceptual Model is relatively stable, system architecture design can commence.

Level 2 Design. A set of product-specific standards and conventions for all aspects of detailed screen/page design is developed (Screen Design Standards), based on any industry and/or corporate standards that have been mandated (e.g., Microsoft Windows, Apple Macintosh, etc.), the data generated in the Requirements Analysis phase, and the product-unique Conceptual Model Design arrived at during Level 1 Design. Screen Design Standards will insure coherence and consistency—the foundations of usability—across the user interface.

The Screen Design Standards (as well as the Conceptual Model Design) are applied to design the detailed user interface to selected subsets of product functionality. This design is implemented as a live prototype (Screen Design Standards Prototyping). An evaluation technique such as formal usability testing is carried out on the Screen Design Standards prototype (Iterative Screen Design Standards Evaluation), and then redesign/reevaluate iterations are performed to refine and validate a robust set of Screen Design Standards. Iterations are continued until all major usability "bugs" are eliminated and usability goals seem within reach.

At the end of the design/evaluate iterations in Design Levels 1 and 2, you have a validated and stabilized Conceptual Model Design, and a validated and stabilized set of standards and conventions for all aspects of detailed Screen Design. These are captured in the document called the product Style Guide, which already documents the results of requirements analysis tasks (Style Guide). During Detailed User Interface Design, following the Conceptual Model Design and Screen Design Standards in the product Style Guide will insure quality, coherence, and consistency, the foundations of usability.

Level 3 Design. Detailed design of the user interface to the complete set of product functionality is carried out based on the refined and validated Conceptual Model Design and Screen Design Standards documented in the product Style Guide (Detailed User Interface Design). This design then drives product design and development. A technique such as formal usability testing is continued during product development to expand evaluation to previously un-assessed subsets of functionality and categories of users, and also to continue to refine the user interface and validate it against usability goals (Iterative Detailed User Interface Design Evaluation).

Phase Three: Installation

After the product has been installed and in production for some time, feedback is gathered to feed into enhancement design, design of new releases, and/or design of new but related products (User Feedback).

Overview of Unique Aspects of Usability Engineering for Cross-Cultural User Interfaces

Various aspects of The Usability Engineering Lifecycle, as described earlier, will need to be modified or added to, and may be more difficult and expensive, when designing for a cross-cultural audience. There will always be a variety of approaches possible to addressing the unique aspects of including usability engineering in a cross-cultural project. Dray and Siegel (chap. 7, this volume) address many of these alternative approaches. Here, we just refer to them briefly to provide context for this chapter on cost justification.

In the Requirements Analysis phase, fundamental and application-independent cross-cultural differences in design, such as use of color, symbols and icons, date and currency formats (referred to extensively in other chapters in this volume) will need to be addressed. The good news here is that most likely all of this information about a culture (i.e., language issues, formatting conventions, use of color, etc.) can probably be accurately learned from a single native individual who also speaks very fluent American English (or whatever is the language of the designing organization). It may be possible, and would certainly be desirable, to hire such a person to be on site with the project team full time. Of course, if there are multiple target cultures, multiple natives must be hired. Another alternative here is to find a firm that specializes in researching and providing this type of cross-cultural information and partnering with them. Still another approach is to simply assume that when in a target country doing the application-unique types of requirements analyses that are a normal part of The Usability Engineering Lifecycle (i.e., User Profile and Contextual Task Analyses), your usability engineer will be attuned to these more general cultural issues, and as a part of other activities, pick up these differences. Depending on which approach is chosen, different costs will be incurred. Even though gathering this type of requirements data is necessary and must be planned for, strictly speaking, it need not be included in the cost justification of a usability engineering program, because as explained earlier, it is really a basic prerequisite for designing for another culture.

While a User Profile questionnaire can be distributed via a web site or e-mail, it may need to be translated into the target language, and recruiting appropriate participants to respond may be more complex than in one's native country. It might be most effective to hire a firm in each target country to recruit questionnaire participants, and communicate the sampling requirements to them. The services of a translator may be necessary to communicate the recruiting task to the local firm.

An alternative to an end user questionnaire for user profiling is to conduct interviews with representatives of target users (e.g., managers) in each country and ask them to characterize the target user population, although just as when using this technique in one's native country, this sort of second hand data is always less reliable. If you do this, you may still need to plan on translation costs to conduct these interviews.

Task Analyses will be considerably more costly and difficult in other countries. One really needs access to actual end users to conduct an effective Task Analysis. One approach is for someone in the designing organization to travel to each of the target countries and, if necessary, employ a translator to both translate all task analysis materials beforehand and translate in real time during the Task Analysis sessions with potential users. Time should also be planned to coordinate with and train the translators regarding what is important to pick up and what is not. You would need to factor in travel time as well, as it may not only take much longer for your usability engineer to get to the target country than to travel within the United States (or whatever is the home country), but there may be a day or two required to recover from jet lag and the time zone change as well.

Alternatively, locals skilled in Task Analysis techniques might be hired in the target countries and communicated with effectively, which may entail translation costs and long distance phone expenses.

In addition, it may take more time to conduct Task Analyses in countries other than the United States. For example, in some countries you might not be able to get people to participate during work hours. In others, you might not get them on weekends. If someone is traveling to the target country to conduct task analysis interviews, you should thus probably plan more elapsed travel time to get the interviews completed than it would take in the United States. Also, in some cultures no-shows may be much more likely than in the United States, especially if they must come to some other facility away from work, so you may have to schedule as many as two to three users for every data point you hope to get, and correspondingly plan the additional time to run sessions with whoever does show up.

In the Design/Testing/Development phase, during design activities it would be very helpful to have a native member of each target country on the team. Planning needs to be done in this phase to build a global architecture that can be most efficiently translated into each local target country. In later stages of design, design will need to be localized for each target country.

Usability tests should be conducted with native end users. One approach is to hire locals in each target country skilled in usability testing and communicate effectively with them, which may entail translation costs and long distance phone expenses. Alternatively, you might send someone in the design organization to the target countries and hire translators to work with them. Just as with performing Task Analyses, it may take more travel time, as well as more time in-country to collect enough data points in a test, as the rate of no-shows may be higher than it would be in one's own country, and it may be harder to schedule test users.

In either case, you need to recruit appropriate users for testing, and costs similar to those described earlier in this chapter for recruiting for User Profiles and Task Analyses need to be factored in.

An alternative is hiring skilled locals to perform heuristic evaluations of prototype designs. In this case, translation costs may be incurred to translate the report into the native language of the development organization.

Many of the same issues arise in the Installation phase where usability feedback is desired after implementation. Users are remote, and there is additional time and expense involved in soliciting their direct feedback. Again, alternatives are hiring skilled locals, or having someone from the design organization travel and work with a translator. Some feedback techniques can be conducted remotely, using the Web or e-mail, and again, for these techniques it might be best to hire a local firm to recruit appropriate participants based on sampling specifications. Again, there may be additional costs for various translation services, and increased travel time and in-country time relative to within one's own country.

FRAMEWORK FOR COST JUSTIFICATION

This rest of this chapter briefly presents a general approach for cost-justifying usability engineering efforts for software development projects. The subject of cost-justifying usability engineering in general is covered extensively in a book called *Cost-Justifying Usability* (Bias & Mayhew, 1994.) In particular, chapter 2 in that book elaborates, in much more detail, on the simple framework offered in this chapter. Other chapters in that book offer further discussion on the topic and case studies. Readers interested in learning more about how, when, and why to cost-justify usability engineering efforts, are referred to that book for more detail than can be provided here. (Also see Bias et al., 2003; Mayhew & Bias, 2003) The purpose of this chapter is to provide an overview of how that more general cost justification technique can be applied to cross-cultural development projects in particular. To that end, a simplified overview is provided, followed by some sample hypothetical analyses for cross-cultural development projects.

Purposes for Cost Justification

There are two possible purposes for conducting a cost–benefit analysis of usability engineering efforts for software development projects: (a) to "sell" usability engineering in general, and (b) to plan for a usability engineering program for a particular development project

Very general cost–benefit analyses of hypothetical usability engineering efforts can be prepared as a strategy to win general support for trying out usability engineering techniques in a development organization. When an organization has no experience with usability engineering, cost–benefit analyses can help to win resources to experiment with it for the first time.

In organizations more mature with respect to usability engineering, cost–benefit analyses can be used to plan an optimal usability engineering effort for a particular development project.

To help settle on a final usability engineering plan for a specific development project, you could start out by calculating the costs of the most aggressive usability engineering program that you might like to implement, including rigorous *techniques* for all lifecycle *tasks*. Then you could estimate benefits, using *very conservative* estimates of benefits. If benefits outweigh costs dramatically, as they usually do when critical parameters are favorable, then you could easily make a good argument for even the most aggressive usability engineering program, because only the most conservative claims concerning potential benefits have been made and, as such, can be defended easily. In fact, you can then go back and redo the benefits estimates using more aggressive, yet still realistic, benefit assumptions, and show that in all likelihood an even more dramatic overall benefit will be realized, even from a significant investment in usability engineering.

If, however, benefits and costs in the initial analysis seem to match up fairly closely, then you will want to scale back your initial, aggressive usability engineering program, maybe even to just a bare-bones plan. Perhaps you should plan to do only a quick and dirty user profile by interviewing user management, a very quick and dirty task analysis consisting of just a few rounds of contextual observations/interviews with users, and then do just one iterative cycle of usability testing on a complete detailed design, to catch major flaws and be sure you have achieved the predicted benefits. You are still likely to realize the very conservative assumptions made concerning benefits with just these minimal usability techniques, and thus you could predict with confidence a healthy return on investment (ROI) from a more minimal approach to usability engineering. This is probably wiser in the long run than making overly optimistic claims concerning potential benefits, spending a lot to achieve them, and then perhaps *not* achieving them and losing credibility.

Thus, you can use a cost–benefit analysis in this fashion, to develop a sensible usability engineering effort for a cross-cultural development project, likely to pay off as predicted.

When an organization is first experimenting with usability engineering techniques, and is still skeptical about their value, it is wise to make very conservative cost–benefit arguments, based on a relatively low cost usability engineering effort and very modest estimates of benefits, and then to show after the fact that much larger benefits were in fact realized. Once an organization has several positive experiences investing in usability engineering, it will be more receptive to more aggressive proposals for usability engineering programs, and to more optimistic benefits estimates in the cost–benefit analyses that argue for them. (See Bias & Mayhew, 1994, chap. 2, for additional concrete examples of how to use cost–benefit analyses to tailor usability engineering programs.)

General Approach

To cost justify any usability engineering plan, cross-cultural or not, you simply adapt a very generic and widely used cost–benefit analysis technique. Having laid out a detailed usability project plan based on lifecycle tasks (see earlier section titled "Overview of a Generic Usability Engineering Lifecycle," and also Mayhew, 1999), it is a fairly straightforward matter to calculate the costs of that plan. Then you need to calculate the benefits. This is a little trickier and where the adaptation of the generic analysis comes into play. Then, you simply compare costs to benefits to find out if, and to what extent, the benefits outweigh the costs. If they do to a satisfactory extent, then you have cost justified the planned effort.

More specifically, first a usability engineering plan is laid out. The plan specifies particular techniques to employ for each lifecycle task, breaks the techniques down into steps, and specifies the personnel hours and equipment and travel costs for each step. The cost of each task is then calculated by multiplying the total number of hours for each type of personnel by their effective hourly wage (fully loaded, as we explain later) and adding in any equipment, travel, and other costs. Then, the costs from all tasks are summed to arrive at a total cost for the plan.

Next, overall benefits are predicted by selecting relevant benefit categories, calculating expected benefits by plugging project specific parameters and assumptions in to benefit formulas, and summing benefits across categories.

The list of possible benefits to consider is long, as usability engineering leads to tangible benefits to all concerned. The development team realizes savings, as problems are identified early when they are cheap to fix. The customer support team realizes a reduced call support burden. More usable e-commerce Web sites have higher buy-to-look ratios, a lower rate of abandoned shopping carts, and increased return visits. More usable applications have fewer errors and higher user productivity. Training costs are lower on more usable applications.

At this point, to simplify the discussion of cost justification for cross-cultural software, we limit the discussion to web sites. It should be noted, however, that similar analyses can be applied to shrink-wrapped software, and software developed for internal use by specific companies. The benefit categories might be slightly different in these cases, but the overall analyses are highly analogous. Other sources provide the details of analyses in these other cases (Bias & Mayhew, 1994; Bias et al., 2003), and could be adapted with the details of cross-cultural usability engineering provided in this chapter.

The potential benefit categories selected in a particular cost–benefit analysis depends on the basic business model for the web site. Sample benefit categories potentially relevant to *e-commerce* sites include: (a) increased buy-to-look ratios, (b) decreased abandoned shopping carts, (c) increased return visits, (d) decreased costs of other sales channels, (e) decreased usage of "Call Back" button, and (f) decreased development costs.

Sample benefit categories potentially relevant to sites based on an *advertising* business model include: (a) increased number of visitors, (b) increased number of return visitors, (c) increased length of visits, (d) increased click-through on ads, (e) decreased failed searches, and (f) decreased development costs.

Sample benefit categories potentially relevant to *product information* sites include: (a) increased sales leads, and (b) decreased development costs.

Sample benefit categories potentially relevant to *customer service* sites include: (a) decreased costs of traditional customer service channels, and (b) decreased development costs.

Sample benefit categories potentially relevant to *intranet* sites supporting internal business users include: (a) decreased user training costs, (b) increased user productivity, (c) decreased user errors, (d) decreased user support, and (e) decreased development costs.

Note that the relevant benefit categories for different types of sites vary somewhat. Thus, in a cost–benefit analysis, one wants to focus attention on the potential benefits that are *of most relevance to the bottom line business goals for the site*.

Also note that these benefits represent just a sample of those that might be relevant in these types of sites, and does not address other possible benefits of usability in other types of sites. Others might be included as appropriate, given the business goals of the site sponsor and the primary concerns of the audience, and could be calculated in a similar fashion as these are (see next section).

Finally, overall benefits are compared to overall costs to see if, and to what extent, the overall usability engineering plan is justified.

SAMPLE COST–BENEFIT ANALYSES FOR CROSS-CULTURAL USABILITY ENGINEERING PROJECTS

As far back as 1991, five of the six major U.S. computer vendors were bringing in over 50% of their income from international sales (Russo & Boor, 1993). Smaller countries had an even larger percentage of their sales come from outside their own country (Nielsen, 1990a). In the decade plus since then, the Internet and the World Wide Web have made it even easier to market and sell in other countries. However, understanding cross-cultural differences, and incorporating them into both products and marketing channels like the Web, has a profound impact on the success of international marketing efforts, just as incorporating an understanding of domestic users and their tasks into products and marketing channels impacts overall success and ROI in one's own country.

Consider a hypothetical usability engineering plan in the context of two different cross-cultural web development projects, and then see how you could conduct cost–benefit analyses of that plan for each project.

A Cross-Cultural E-Commerce Web Site

Imagine a web development organization is planning to redesign an existing e-commerce site, which is not producing the ROI hoped for. Traffic and sales statistics are available from the current site. In particular, assume the original site was designed in English for a U.S. audience, but is being used by multiple other cultures, and is not performing as well as hoped in any country, but especially in the international countries. Part of the goal of the redesign is to tailor the site for use in a small number of target countries. The usability engineer prepares a proposed usability engineering project plan to be conducted for *each* target country (including the country of the development organization), and then performs a cost–benefit analysis of that plan.

First, the final results of cost–benefit analyses for a "base" country (in this example, the United States) and *one* international country are presented in the rest of this section. Then, in the sections that follow, the derivations of those final results are described. The same type of analysis would apply to *each* international country, although the numbers would vary from country to country depending on costs in individual countries. It should be noted that the cost for usability engineering programs would probably be less for a set of countries than the sum of the costs computed separately for each country, as certain parts of the program could be generalized or "amortized" across countries. For example, creating usability testing materials would likely take the most time in the first country, but be a simple matter of translation for other countries. However, for simplicity's sake, here we assume and present the analysis for just one international country.

Table 8.1 shows the overall calculation of the *cost* of a usability engineering plan for a project including a base country (in this example, the United States) and one international country (imagine a country in, say, South America, where the cost of living and pay scales are significantly lower than in the United States). In Table 8.1, the first column identifies the overall project phases. The second column identifies which Usability Engineering Lifecycle tasks (see "Overview" section at chapter beginning) are planned in each phase. This column also breaks down each task according to general categories of costs, first for the base country (in this example, the United States), and then for the international country.

The third, fourth, fifth, sixth, and seventh columns identify the number of hours of time of different types of professionals that are required to complete each task. The last column then summarizes the total cost of each cost category within a task. Note that values in the cells representing hours and costs for the international country are expressed as *differences* relative to the base country, referred to in Table 8.1 as "International Delta." For example, under User Profile, you will see that relative to the base country, 12 more hours of the usability engineer's time (beyond the 80 hours required for doing a User Profile at home) are expected to be required.

Grand cost totals for the whole plan are given at the bottom of the table. All dollar amounts are given in the currency of the base country, in this example, U.S. Dollars.

TABLE 8.1
Cost of Usability Engineering Plan

PHASE	TASK	Usability Engineer Hours @ $150	Developer Hours @ $150	User Hours @ $40	User Hours @ $20	Language/Culture Consultant Hours @ $75	TOTAL COST
Requirements Analysis	**User Profile**						
	BASE						
	Hours	80		12.5			$12,500
	Recruitment Fees						$2,000
	INTERNATIONAL DELTA						
	Hours	12		– 12.5	12.5	12	+ $2,450
	Recruitment Fees						– $1,000
	Materials Translation						+ $1,050
	Task Analysis						
	BASE						
	Hours	104		30			$16,800
	Recruitment Fees						$800
	Travel/Phone Expenses						$3,275
	INTERNATIONAL DELTA						
	Hours	80		– 30	30	32	+ $13,800

(Continued on next page)

225

TABLE 8.1 (continued)

PHASE	TASK	Usability Engineer Hours @ $150	Developer Hours @ $150	User Hours @ $40	User Hours @ $20	Language/ Culture Consultant Hours @ $75	TOTAL COST
	Recruitment Fees						− $400
	Materials Translation						+ $300
	Local Facilitator						+ $2,500
	Simultaneous Translation						+ $2,000
	Visa						+ $180
	Travel/Phone Expenses						+ $1,575
Platform Constraints							
	BASE						
	Hours	8	8				$2,400
Usability Goal Setting							
	BASE						
	Hours	20					$3,000
	INTERNATIONAL DELTA						
	Hours					+8	+ $600

Design/Test/Develop			
Information Architecture			
BASE			
Hours	60		$9,000
INTERNATIONAL DELTA			
Hours	16	16	+ $3,600
Conceptual Model Design			
BASE			
Hours	60		$9,000
INTERNATIONAL DELTA			
Hours	16	40	+ $5,400
Screen Design Standards			
BASE			
Hours	60		$9,000
INTERNATIONAL DELTA			
Hours	16	40	+ $5,400
Live Prototype Development			
BASE			
Hours	20	120	$21,000
INTERNATIONAL DELTA			
Hours	16	40	+ $5,400

(Continued on next page)

227

TABLE 8.1 (continued)

PHASE	TASK	Usability Engineer Hours @ $150	Developer Hours @ $150	User Hours @ $40	User Hours @ $20	Language/ Culture Consultant Hours @ $75	TOTAL COST
Usability Test							
	BASE						
	Hours	104		16			$16,240
	Recruitment Fees						$400
	Travel/Phone Expenses						$2,775
	INTERNATIONAL DELTA						
	Hours	56		−16	+16	+16	$9,280
	Recruitment Fees						−$200
	Materials Translation						+$3,000
	Local Facilitator						+$2,500
	Simultaneous Translation						$2,000
	Visa						+$180
	Travel/Phone Expenses						+$1,350
Redesign							
	BASE						
	Hours	40					$6,000
	INTERNATIONAL DELTA						
	Hours	16				16	$3,600

Detailed User Interface Design					
BASE					
Hours	60				$9,000
INTERNATIONAL DELTA					
Hours	16			16	+ $3,600
Live Prototype Development					
BASE					
Hours	20	80			$15,000
INTERNATIONAL DELTA					
Hours	16			16	+ $3,600
Usability Test					
BASE					
Hours	84		16		$13,240
Recruitment Fees					$400
Travel/Phone Expenses					$2,775
INTERNATIONAL DELTA					
Hours	40		– 16	16	$8,080
Recruitment Fees					– $200
Materials Translation					+ $3,000
Local Facilitator					+ $2,500
Simultaneous Translation					+ $2,000

(Continued on next page)

229

TABLE 8.1 (continued)

PHASE	TASK	Usability Engineer Hours @ $150	Developer Hours @ $150	User Hours @ $40	User Hours @ $20	Language/ Culture Consultant Hours @ $75	TOTAL COST
	Visa						+ $180
	Travel/Phone Expenses						+ $1,350
Redesign							
	BASE						
	Hours	40					$6,000
	INTERNATIONAL DELTA						
	Hours	16				16	+ $3,600
	TOTAL BASE	760	208	74.5	0	0	$160,605
	INTERNATIONAL DELTA	316	0	− 74.5	+ 74.5	+ 284	+ $92,275
	TOTAL INTERNATIONAL	1,076	208	0	74.5	284	$252,880

Next, the project usability engineer estimates that the usability engineering plans described in Fig. 8.1 will produce new site designs with particular expected benefits every month in the two countries, as shown in Table 8.2.

Comparing these benefits and costs, the project usability engineer argues that the proposed usability engineering plan at home will likely pay for itself in the first five months after launch:

Base Country

Benefits per month =	$33,854
One-time cost =	$160,605
PAY-OFF PERIOD =	4.74 months

Given the higher costs for the international usability engineering program, plus the different parameters and predictions of benefits, the project usability engineer then argues that the proposed international usability engineering plan will likely pay for itself within the first seven months after launch:

International Country

Benefits per month =	$40,625
One-time cost =	$252,880
PAY-OFF PERIOD =	6.23 months

As the new sites are expected to have a lifetime of something much longer than 7 months, the project usability engineer expects the plans to be approved *for these two countries in particular*, based on this cost justification.

TABLE 8.2
Expected Monthly Benefits for an e-Business Site

BENEFIT CATEGORY	VALUE in Base Country— PER MONTH	VALUE in International Country— PER MONTH
Decreased abandoned shopping carts	$15,625	$18,750
Increased buy-to-look ratio	$15,625	$18,750
Decreased use of "Call Back" button	$2,604	$3,125
TOTAL MONTHLY BENEFITS	$33,854	$40,625

Note that the analyses offered here do not consider the time value of money. That is, the money for the costs is spent at some point in time, whereas the benefits come later in time. Also, if the money was NOT spent on the costs of usability engineering, but instead was invested, this money would likely increase in value. This investment could be compared to the benefits of investing in the usability engineering program. Furthermore, both costs and benefits are somewhat simplified and not exhaustive. Some expenses, but also some benefits, may have been overlooked. Usually, the benefits of usability are so robust that these more sophisticated and complex financial considerations and details are not necessary. However, if needed, calculations based on the time value of money are presented in Bias et al. (2003), and also in Bias and Mayhew (1994).

Next is a step-by-step description of how the project usability engineer arrived at the final results described in this section.

Start With a Usability Engineering Plan

If it has not already been done, this is the first step in conducting a cost–benefit analysis. The usability engineering plan identifies which Usability Engineering Lifecycle tasks and techniques will be employed, and breaks them down into required staff, hours, and other expenses. Costs can then be computed for these tasks in the next two steps. The sample usability engineering plan used in these examples and laid out in Table 8.1 includes most lifecycle tasks and fairly rigorous techniques for each task. It assumes that three tasks (i.e., the task analysis and two rounds of usability testing) require travel, but that the rest of the work can be done in the base country location of the development organization.

Establish Analysis Parameters

Most of the calculations for both planned costs and estimated benefits are based on project-specific parameters. These should be established and documented before proceeding with the analysis. Sample parameters are given below for our hypothetical project.

Base (U.S.) Parameters:

- The site is an e-business site (vs. a product information site, or a site based on an advertising model).
- The *current site* gets an average of 125,000 visitors per month in the base country (vs., for example, 500,000 users or 10,000 users).
- The current buy-to-look ratio in the base country is 2%.
- The current average profit margin on each online purchase is $25.

- The current rate of usage of the "Call Back" button (serviced by a local customer service organization) is 2%.
- Average time to service each usage of the "Call Back" button is 5 minutes.
- Users are paid to participate in usability engineer tasks at an hourly rate of $40.
- Customer support fully loaded hourly wage is $50.
- Usability Engineer and Developer fully loaded hourly wage is $150.

International (for one hypothetical country) Parameters:

- The site is an e-business site (vs. a product information site, or a site based on an advertising model).
- The *current site* gets an average of 75,000 visitors per month in the international country.
- The current buy-to-look ratio in the international country is 1%.
- The current average profit margin on each online purchase is $25.
- The current rate of usage of the "Call Back" button (serviced by a local customer service organization) is 4%.
- Average time to service each usage of the "Call Back" button is 10 minutes.
- Users are paid to participate in usability engineering tasks at an hourly rate of $20.
- Customer support fully loaded hourly wage is $25.
- Usability Engineer and Developer fully loaded hourly wage is $150.
- Language/culture consultant is a graduate student living in the location of the development organization in the base country, paid a flat rate of $75 per hour.

Fully loaded hourly wages are calculated by adding together the costs of salary, benefits, office space, equipment, and other facilities for a type of personnel, and dividing this by the number of hours worked each year by that personnel type. If outside consultants or contractors are used, their simple hourly rate would apply.

The hourly rate for users used in this cost estimate is $40 in the base country and $20 in the international country. This is not based on a typical user's fully loaded hourly wage at his or her job, which it would be in the case of a cost justification of traditional software development for internal users or intranet development for internal users. Instead, it is premised on the assumption that test users in the case of an e-commerce web site have to be recruited from the general public to participate in usability engineering tasks/techniques, and that they are paid at a rate of $40 (or $20) an hour for their participation.

The fully loaded hourly rate for usability engineering staff is based roughly on a typical current salary and benefits for a senior-level internal usability engineer in the United States. The hourly rate of developers was similarly estimated.

It should be emphasized that when using the general cost–benefit analysis technique illustrated here, the particular parameter values used in the sample analyses should not be assumed. The particular parameter values of *your* project and organization and *your* base and international countries should be substituted for those given. They will almost certainly be different from the parameters used in this example. In particular, professionals hired in different countries may have very different rates in the designing organization's currency.

Calculate the Costs of the Usability Engineering Plan

Given a usability engineering plan, the number of hours required of each necessary type of staff can be estimated for each task/technique in the plan. Once you can estimate the number of hours required for each task/technique, you then simply multiply the total number of hours of each type of staff required for each task by the fully loaded hourly wage of that type of staff, and then sum across the different staff types. Additional costs, such as equipment, supplies, travel expenses and special services, should also be estimated and added in for each task. This is how the task costs in Table 8.1 were calculated.

The numbers of hours estimated for each task/technique in the cost summary table were not pulled out of a hat—they are hypothetical, but based on many years of experience. For example, the 104 hours estimated to conduct the first usability test in the base country (in this case, the United States) was derived as follows:

- Design/develop test materials   40 hrs
- 2 days' travel   16
- Run test/collect data (3 days, 8 test users)   24
- Summarize/interpret data; draw conclusions   8
- Document results   16
- TOTAL   104

Similar sample breakdowns of the steps required in each Usability Engineering Lifecycle task/technique can be found in Mayhew (1999) and in Bias and Mayhew (1994).

The cost differences for the international usability engineering program indicated in Table 8.1 are also hypothetical, but again, are extrapolated from the actual experience of some of our colleagues conducting such projects. For example, in the case of the same first usability test, the additional 56 hours of the usability engineer's time is based on a number of anticipated factors:

- More time spent in preparation for the testing due to the complexity of making international travel and support staff arrangements.
- More travel time (actual in-transit time, plus a weekend stay-over required to be on site the required number of weekdays).
- The assumption of more no shows in the target country, which means more users must be scheduled in order to meet the goal of eight data points, which in turn means the usability engineer must plan to spend more time at the testing site.
- Time spent training the local facilitator before testing begins.
- Time spent simply picking up basic cross-cultural differences (referred to earlier in this chapter). It is assumed that basic data were provided, but that additional data will be pursued and picked up during testing.

Also note that there are cost categories in Table 8.1 that are given as flat fees, rather than broken down into staff types, numbers of hours of each staff type, and hourly rates. For example, flat fees are given for services such as Recruitment of Users, Local Facilitator, and Simultaneous Translation. These dollar amounts are extrapolated from the actual experience of colleagues in countries where service providers quoted a flat fee for the service agreed to, rather than an hourly rate.

The difference in travel expenses for the base and international countries is primarily due to the expectation of higher airfares for international travel. Whereas more travel days are required, both in-transit and on-site, in order to run the same number of test users as in the base country, it is expected that hotel, meal, and rental car costs will be less in the international country, making overall travel expenses comparable in the two countries.

Remember that both the hourly wage figures and the predicted hours and expenses per cost category for each task used to generate the sample analyses here are just hypothetical examples based on specific experiences. Again, you have to use the actual costs and time of personnel and other expenses in your country and in your target country(s), in order to carry out your own analysis.

Select Relevant Benefit Categories

Since this is an *e-business site*, only certain benefit categories are of relevance to the business goals of this site. For another kind of site, different benefit categories would be selected.

In this hypothetical case, the project usability engineer decides to include the following benefits: (a) increased buy-to-look ratio, (b) decreased abandoned shopping carts, and (c) decreased use of "Call Back" button.

These categories were selected because the usability engineer knew these would be of most relevance to the analysis audience, the business sponsors of the site. There may be other very real potential benefits of the usability engineering plan, but the usability engineer chose just these for simplicity and to make a conservative estimate of benefits.

As compared to the existing site design, the usability engineer anticipated that in the course of redesign, the usability engineering effort would decrease abandoned shopping carts by insuring that the checkout process is clear, efficient, provides all the right information at the right time, does not violate any cultural expectations or values, and does not bother users with tedious data entry of information they do not want or need to provide. The usability engineer expected to improve the buy-to-look ratio by insuring that the right product information is contained in the site, that navigation to find products is efficient and always successful, and that no cultural blunders are made in terms of product names, site use of colors, and so forth. The usability engineer expected to decrease the use of the "Call Back" button by making the information architecture match user expectations by designing and validating a clear conceptual model, so that navigation of, and interactions with, the site are intuitively obvious, and by insuring that language translation is not misleading or confusing. Accomplishing all these things depends on conducting the requirements analysis, design and testing activities in the proposed plan with an eye towards uncovering and addressing unique requirements in the international country, as well as on applying general user interface design expertise.

Quantify/Estimate Benefits

Next, the project usability engineer estimates the magnitude of each benefit that would be realized (compared to the current site, which is being redesigned) *if* the usability engineering plan (with its associated costs) were implemented. Thus, for example, she or he estimates how much *higher* the buy-to-look ratio would be on the site if it were reengineered for usability as compared to the existing site.

To estimate each benefit, you must choose a unit of measurement for the benefit, such as the average purchase profit margin in the case of the Increased Buy-to-Look Ratio benefit, or the average cost of customer support time spent servicing each usage of the "Call Back" button, in the case of the Reduced Use of the "Call Back" Button benefit. Then—and this is the tricky part—you must make an assumption concerning the magnitude of the benefit for each unit of measurement, for example, a 1% increase in buy-to-look ratio, or a 1% decrease in the usage rate of the "Call Back" button. (Tips on how to make these key assumptions are discussed later.) Finally, you will calculate the total benefit in each category based on the unit of measurement, key parameters, and your assumptions about magnitudes of benefit. When the unit of measurement is time, benefits can be expressed first in units of time and then converted to dollars, given the value of time.

Remember that our hypothetical project involves development of an e-commerce site. Based in part on in-house experience, our project usability engineer makes the following key assumptions:

- Buy-to-look ratio will increase by .5% of total visitors in the United States and by 1% in the international country.
- Abandoned shopping carts will decrease by .5% of total visitors in the United States and by 1% in the international country.
- Usage of the "Call Back" button will decrease by .5% of total visitors in the United States and by 1% in the international country.

Note that the assumption is that there will be greater benefits in the international country than in the base country. Remember that currently the international country is using the site designed in English for a U.S. audience. Any general usability problems are thus compounded by the fact that the site is also not in the user's native language, nor does it take into account other cross-cultural differences, such as preferred measurement units, culture-appropriate imagery, culture-appropriate use of color, and currency differences. Thus, due to decreasing the general "alien-ness" of the site for the international audience, as well as improving the general usability of the site, greater benefits are predicted.

Also note that very conservative assumptions regarding predicted benefits are made, in spite of the aggressiveness and thoroughness of the usability engineering project plan. A 0.5% increase in the buy-to-look ratio in the base country is not much. However, making conservative benefits estimates in a cost justification analysis such as this is always wise: If you can show an overall benefit, even when your costs are high and your claims regarding expected benefits are very conservative, then you have a compelling argument for your plan.

Based on these key assumptions, the project usability engineer then calculates benefits in each of the selected benefit categories as follows:

Increased Buy-to-Look Ratio:
  Base Country:
    .5% more visitors will decide to buy, and will successfully make a purchase each month
    125,000 visitors per month × .5% = 625 more purchases
    625 purchases @ profit margin of $25 = **$15,625 per month**
  International Country:
    1% more visitors will decide to buy, and will successfully make a purchase each month

75,000 visitors per month $\times$ 1% = 750 more purchases
750 purchases @ profit margin of $25 = **$18,750 per month**

Decreased Abandoned Shopping Carts:
Base Country:
.5% more visitors who would have decided to buy anyway will now
complete checkout successfully and make a purchase each
month
125,000 visitors per month $\times$ .5% = 625 more purchases
625 purchases @ profit margin of $25 = **$15,625 per month**
International Country:
1% more visitors will decide to buy, and will successfully make a
purchase each month
75,000 visitors per month $\times$ 1% = 750 more purchases
750 purchases @ profit margin of $25 = **$18,750 per month**

Decreased Usage of "Call Back" Button:
Base Country:
.5% fewer visitors will need to use the "Call Back" button each
month
125,000 visitors per month $\times$ 5% = 625 fewer calls
625 calls at 5 minutes each = 52.08 hours
52.08 hours @ $50 = **$2,604 per month**
International Country:
1% fewer visitors will need to use the "Call Back" button each
month
75,000 visitors per month $\times$ 1% = 750 fewer calls
750 calls at 10 minutes each 125 hours
125 hours @ $25 = **$3,125 per month**

This is how the benefit predictions summarized in Table 8.2 were made.

The usability engineer based the assumptions regarding benefits in the base country on statistics available in the literature, such as those presented in Mayhew and Bias (1994). In particular, she or he began with the often quoted average e-commerce web site buy-to-look ratio of 2–3%. The usability engineer then based the assumption that this rate could be improved by a minimum of 1% (.5% from improving the product search process and .5% from improving the checkout process) through usability engineering techniques, on statistics offered as average by a Forrester report called "Get ROI From Design" (Souza, 2001). This report suggested that it would be typical for as much as 5% of online shoppers to fail to find the product and offer they are looking for (other statistics suggest as much as 45% may experience this problem), and for over 50% of shoppers that do find a product they would like to buy to bail out during checkout due to confusion over forced registration, being asked to give a credit card number before being shown total costs, and because they are not convinced this is the best available offer. The assumption of reduced usage of the "Call Back" button by .5% was based on statistics cited earlier suggesting as much as 20% of site users typically call in to get more information. Most of us have experienced these problems and would have little argument with the idea that they are typical. Given these statistics, the assumptions made for the base country seem very modest indeed. The slightly greater benefits assumptions made for the international country were based on the expectation that decreasing the general "alien-ness" of the site for the international audience would provide benefits beyond improving the general usability of the site.

The basic assumption of a cost–benefit analysis of a usability engineering plan is that the improved user interfaces that are achieved through usability engineering techniques result in such tangible, measurable benefits as those calculated in this hypothetical example.

The audience for the analysis is asked to accept these *assumptions* of certain estimated, quantified benefits as *reasonable and likely minimum benefits*, rather than as precise, proven, guaranteed benefits. Proof simply does not exist for each specific web site that an optimal user interface provides some specific, reliable advantage over some other user interface that would, or has, resulted in the absence of a usability engineering plan.

How can you generate, and convince your audience to accept, the inherent assumptions in the benefits you estimate in a given cost–benefit analysis? First, it should be pointed out that *any* cost–benefit analysis for *any* purpose must make certain assumptions that are really only predictions of the likely outcome of investments of various sorts. The whole point of a cost–benefit analysis is to try to evaluate in advance, in a situation in which there is some element of uncertainty, the likelihood that an investment will pay off. The trick is in basing the predictions of uncertainties on a firm foundation of known facts. In the case of a cost–benefit analysis of a planned usability engineering program, there are several foundations on which to build realistic predictions of benefits.

First, there is ample research published that shows measurable and significant performance advantages of specific user interface design alternatives (as compared to other alternatives) under certain circumstances.

Ben Shneiderman (a well-known author in the field of usability engineering) states that "Performance decreases 5% to 10% when designers change the color and position of interface elements. It slows down 25% when developers switch terminology on buttons like 'submit' and 'search'" (Sonderegger, 1998).

In another example, Lee and Bowers (1997) studied the conditions under which the most learning occurred from presented information and found that compared to a control group that had no learning opportunity:

- Hearing spoken text and looking at graphics resulted in 91% more learning.
- Looking at graphics alone—63% more.
- Reading printed text plus looking at graphics—56% more.
- Listening to spoken text, reading text, and looking at graphics—46% more.
- Hearing spoken text plus reading printed text—32% more.
- Reading printed text alone—12% more.
- Hearing spoken text alone—7% more.

Clearly, how information is presented makes a significant difference in what is retained.

The research does not provide simple generic answers to design questions. However, what the research does provide are general ideas of the magnitude of performance differences that can occur between optimal and sub-optimal interface design alternatives. The basic benefit assumptions made in any cost–benefit analysis can thus be generated and defended by referring to the wide body of published research data that exists. From these studies, you can extrapolate to make some reasonable predictions about the order of magnitude of differences you might expect to see in web site user interfaces optimized through usability engineering techniques. (See Bias & Mayhew, 1994, chap. 2, for a review and analysis of some of the literature for the purpose of defending benefit assumptions.)

Besides citing relevant research literature, there are other ways to arrive at and defend your benefit assumptions. Actual case histories of the benefits achieved as a result of applying usability engineering techniques are very useful in helping to defend the benefits assumptions of a particular cost–benefit analysis. A few published case histories exist (e.g., Bias & Mayhew, 1994; Karat, 1989). Wixon and Wilson (1997) and Whiteside Bennett, and Holtzblatt (1988) report that across their experience with many projects over many years, they find that they average an overall performance improvement of about 30% when at least 70% to 80% of the problems they identify during usability testing are addressed by designers.

But even anecdotes are useful. For example, a colleague working at a vendor company once told us that they had compared customer support calls on a product for which they had recently developed and introduced a new, usability-engineered release. Calls to customer support *after* the new release were decreased by 30%. This savings greatly outweighed the cost of the usability engineering effort. (For another, even more dramatic anecdote, see the example cited by Nielsen, 1993, p. 84, regarding an upgrade to a spreadsheet application.)

Finally, experienced usability engineers can draw on their own general experience evaluating and testing software user interfaces, and their specific experience with a particular development organization. Familiarity with typical interface designs from a particular development organization allows the usability engineer to decide how much improvement to expect from applying usability engineering techniques in that organization. If the designers are generally untrained and inexperienced in interface design and typically design poor interfaces, the usability engineer would feel comfortable and justified defending more aggressive benefits claims. On the other hand, if the usability engineer knows the development organization to be quite experienced and effective in interface design, then more conservative estimates of benefits would be appropriate, on the assumption that usability engineering techniques result in fine tuning of the interface, but not radical improvements. The usability engineer can also assess typical interfaces from a given development organization against well-known and accepted design principles, usability test results, and research literature to help defend the assumptions made when estimating benefits.

In general, it is usually wise to make *very conservative* benefit assumptions. This is because any cost–benefit analysis has an intended audience, who must be convinced that benefits will likely outweigh costs. Assumptions that are very conservative are less likely to be challenged by the relevant audience, thus increasing the likelihood of acceptance of the analysis conclusions. In addition, conservative benefits assumptions help to manage expectations. It is always better to achieve a greater benefit than was predicted in the cost–benefit analysis, than to achieve less benefit, even if it still outweighs the costs. Having underestimated benefits will make future cost–benefit analyses more credible and more readily accepted.

When each relevant benefit has been calculated for a common unit of time (e.g., per month or per year), then it is time to add up all benefit category estimates for a benefit total.

Compare Costs to Benefits

Recall that in our hypothetical project, the project usability engineer compared costs to benefits, and this is what was found:

Base Country:

Benefits per month =	$33,854
One-time cost =	$160,605
PAY-OFF PERIOD =	4.74 months

International Country:

Benefits per month =	$40,625
One-time cost =	$252,880
PAY-OFF PERIOD =	6.23 months

Our project usability engineer's initial usability engineering plan for both countries appears to be well justified. It is a fairly aggressive plan (in that it includes all lifecycle tasks and moderate to very rigorous techniques for each task), and the benefit assumptions are fairly conservative. Given the very clear net benefit, it would be wise to stick with this aggressive plan and submit it to project management for approval. In fact, based on the very modest assumptions, it might be well advised to redesign the usability engineering plan to be even more thorough and aggressive because increased benefits that might be realized by a more rigorous approach will likely be more than compensated for. In this case, the usability engineer might consider increasing the level of effort of the requirements analysis tasks, which usually have a high pay-off.

If the estimated pay-off period is long, or in fact if there is no reasonable pay-off period, then it would be wise to go back and rethink the plan, scaling back on the rigorousness of techniques for certain tasks, and even eliminating some tasks (e.g., collapsing the design process from two to one design level, i.e., doing only one usability test) in order to reduce the costs.

A Cross-Cultural Product Information Web Site

This example is based on a hypothetical scenario given in a Forrester report called "Get ROI From Design" (Souza, 2001). It involves an automobile manufacturing company, who has put up a web site that allows customers to get information about the features of the different models of cars they offer, and options available on those cars. It allows users to configure a base model with options of their choice and get sticker price information. Users cannot purchase a car online through this site; it is meant to generate leads and point users to dealerships and salespeople in their area.

Start With a Usability Engineering Plan

In this example, we again start with the same assumed plan as in the e-commerce site example (see Table 8.1).

Establish Analysis Parameters

Sample parameters are given here for this example. Again, we assume there is an existing site with known traffic and sales statistics, and the project involves a redesign.

Base Country Parameters:

- The site is a product information site (vs. an e-business site, or a site based on an advertising model).
- The *current site* gets an average of 500,000 visitors per month (vs., for example, 125,000 visitors or 10,000 visitors).
- Currently 1% of visitors result in a concrete lead.
- Currently 10% of leads generate a sale.
- The profit on a sale averages $300.
- Users are paid to participate in usability engineering tasks at an hourly rate of $40.
- UE and Developer fully loaded hourly wage is $150.

International Country Parameters:

- The site is a product information site (vs. an e-business site, or a site based on an advertising model).
- The *current site* gets an average of 250,000 visitors per month (vs., for example, 125,000 visitors or 10,000 visitors).
- Currently .5% of visitors result in a concrete lead.
- Currently 5% of leads generate a sale.
- The profit on a sale averages $300.
- Users are paid to participate in usability engineering tasks at an hourly rate of $20.
- UE and Developer fully loaded hourly wage is $150.

Calculate the Costs of the Usability Engineering Plan

We can use the same cost calculations as before, shown in Table 8.1.

Select Relevant Benefit Categories

Because this is a *product information site*, only certain benefit categories are of relevance to the business goals of this redesign project. The project usability engineer decides to include only the following benefit category: increased lead generation.

The usability engineer selected this benefit category because it is of most relevance to the audience for the analysis, the business sponsors of the site. There may be other very real potential benefits of the usability engineering plan, but only this one was chosen, for simplicity and to make a conservative estimate of benefits.

As compared to the existing site design, the usability engineer anticipates that in the course of redesign, the usability engineering effort will increase leads by insuring that visitors can find basic information and successfully configure models with options. Accomplishing this depends on conducting the requirements analysis and testing activities in the proposed plan, as well as on applying general user interface design expertise.

Quantify/Estimate Benefits

Next, the project usability engineer estimates the magnitude of the benefit that would be realized (relative to the current site, which is being redesigned) *if* the usability engineering plan (with its associated costs) were implemented. Thus, in this case, she or he estimates how much *higher* the lead generation rate would be on the site if it were reengineered for usability as compared to the existing site. Table 8.3 summarizes these benefits predictions.

The project usability engineer calculated the estimated benefits as follows:

Increased Lead Generation Rate:
   Base Country
      .5% *more* visitors will generate a lead
      500,000 visitors per month × .5% = 2,500 more leads
      10% of these new leads will result in a sale = 250 more sales
      250 more sales @ profit margin of $300 = **$75,000 per month**
   International Country:
      1% *more* visitors will generate a lead
      250,000 visitors per month 1% = 2,500 more leads
      5% of these new leads will result in a sale = 125 more sales
      125 more sales @ profit margin of $300 = **$37,500 per month**

Compare Costs to Benefits

Next, the usability engineer compares benefits and costs to determine the pay-off period:

TABLE 8.3
Expected Monthly Benefits for a Product Information Site

BENEFIT CATEGORY	VALUE in Base Country— PER MONTH	VALUE in International Country— PER MONTH
Increased lead generation	$75,000	$37,500
TOTAL MONTHLY BENEFITS	$75,000	$37,500

Base Country:

Benefits per month =	$75,000
One-time cost =	$160,605
PAY-OFF PERIOD =	2.14 months

International Country:

Benefits per month =	$37,500
One-time cost =	$252,880
PAY-OFF PERIOD =	6.74 months

Again, our project usability engineer's initial usability engineering plan appears to be well justified. It was a fairly aggressive plan, in that it included all lifecycle tasks and moderate to very rigorous techniques for each task, and the benefit assumptions were fairly conservative. Given the relatively short estimated pay-off period and the expectation of a site lifetime of much longer than 7 months, it would be wise to stick with this aggressive plan and submit it to project management for approval. In fact, based on the very modest assumptions regarding increases in lead generation, the project usability engineer might consider redesigning the usability engineering plan to be even more thorough and aggressive because the increased benefits that might be realized by a more rigorous approach will likely be more than compensated for. In this case, it might be advisable to increase the level of effort of the requirements analysis tasks, which usually have a high pay-off.

SUMMARY AND CONCLUSIONS

The previous examples are based on a simple subset of all actual costs and potential benefits, and very simple and basic assumptions regarding the value of money. More complex and sophisticated analyses can be calculated (see Bias & Mayhew, 1994; Bias et al., 2003.) However, usually a simple and straightforward analysis of the type offered in the previous examples is sufficient for the purpose of winning funding for usability engineering investments in general, or planning appropriate usability engineering programs for specific development projects.

One key to applying cost justification to planning usability engineering programs for web development projects is the generation of web traffic and sales statistics that are relevant and useful. In the Forrester Report called "Measuring Web Success" (Schmitt, 1999), it is pointed out that typical web traffic statistics do not really tell you what you want to know. For example, knowing how many visitors to a product information site there were this month, and how many of them viewed each product, does *not* tell you how many customers were satisfied with the information provided. Similarly, knowing how many visitors to an e-commerce site made a purchase this month and what the average purchase price was does *not* tell you why more shoppers did not buy. And, knowing how many visits a customer service site received and how many e-mails were received and responded to by live customer service representatives does *not* tell you how many customer support issues were resolved online. When we have traffic software that can generate more specific and integrated statistics that allow these kinds of inferences, it will become much easier to validate cost justification estimates after the fact, and make future cost–benefit analyses much more credible.

Ideally, says this report, "Managers will have self-service access to relevant information: Human resources will get reports on job seekers, web designers will see detail on failed navigations, and marketing will receive customer retention metrics" (Schmitt, 1999, p. 11). And, "For example, a marketing director trying to determine why look-to-buy ratios plummet after 6:00 pm might discover that more evening users connect through dial-up modems that produce a slower and more frustrating shopping experience" (p. 11).

This report also points out that *cross-channel* statistics that track a given customer are just as important, if not more so, than simple web traffic statistics. For example, if it could be captured that a *particular user* visited a customer service web site, *but then shortly thereafter called the customer service hot line,* this might point to some flaw or omission in the information in the online customer service channel. In addition, if the customer's "tracks" through the web site, as well as the nature of the phone conversation, could also be captured *and integrated,* inferences might be made about exactly how the web site failed to address the customer's need.

This kind of very specific traffic tracking starts to sound very much like a sort of ongoing automated usability test, which could generate data that would point very directly to specific user interface design problems and solutions, making the case for usability engineering even stronger.

One might look at typical web development time frames during the early years of web development, and notice that while the average web development project back then might have taken all of 8 to 12 weeks in total from planning to launch, the usability engineering program laid out in the sample analyses in this chapter would itself take an absolute minimum of 5 months in the base country, and 6 months in the international country. How can it make sense to propose a usability engineering program—most of which would occur during requirements analysis and design of the overall development effort—that takes significantly longer to carry out than the typical whole development effort?

Initially, web sites were functionally very simple compared to most traditional software applications, and so the fact that they typically took 8 to 12 weeks to develop, as compared to months or even years for traditional software applications, made some sense. In addition, very few were explicitly intended to be used internationally. Now, however, web sites and applications have gotten more and more complex and are, in many cases, much like traditional applications that happen to be implemented on a browser platform. In addition, web applications aimed at the international market have gotten much more common. The industry needs to adapt its notion of reasonable, feasible, and effective time frames (and budgets) for developing complex web-based applications for use in multiple other cultures, which simply are not the same as simple content-only web sites intended only for one's own culture. This includes adapting its notion of what kind of usability engineering techniques should be invested in.

In a report by Forrester Research, Inc., called "Scenario Design" (their term for usability engineering), it is pointed out that:

Executives Must Buy Into Realistic Development Time Lines and Budgets

The mad Internet rush of the late 1990's produced the slipshod experiences that we see today. As firms move forward, they must shed their misplaced fascination with first-mover advantage in favor of lasting strategies that lean on quality of experience.

- **Even single-channel initiatives will take eight to 12 months.** The time required to conduct field research, interpret the gathered information, and formulate implementation specs for a new web-based application will take four to six months. To prototype, build, and launch the effort will take another four to six months. This period will lengthen as the number of scenarios involved rises.
- **These projects will cost at least \$1.5 million in outside help.** Firms will turn to eCommerce integrators and user experience specialists for the hard-to-find-experts, technical expertise, and collaborative methodologies required to conduct Scenario Design. Hiring these outside resources can be costly, with run rates from \$150K to \$200K per month. This expenditure is in addition to the cost of internal resources, such as project owners responsible for the effort's overall success and IT resources handling integrations with legacy systems. (Sonderegger, 2000, p. 12)

We agree that 8 to 12 *months* is a more realistic time frame (than 8 to 12 *weeks*) to develop a usable web site or application that will provide a decent ROI—*in one's home country*. Sites which must be localized to multiple other cultures obviously should expect to take longer. And, if this is the overall project time frame, there is ample time in the overall schedule to carry out a usability engineering program such as the one laid out earlier in this chapter. The total cost of the sample usability engineering program offered here is also well within Sonderegger's estimate of the costs of "hard-to-find-experts," which include "user experience specialists." The sample cost justification analyses offered here, as well as others offered by a Forrester report called "Get ROI From Design" (Souza, 2001), suggest that it is usually fairly easy to justify a significant time and money investment in usability engineering during the development of web sites. And the framework and examples presented in this chapter should help you demonstrate that this is the case for your international web project.

ACKNOWLEDGMENTS

Portions of this chapter are excerpted or adapted from *The Usability Engineering Lifecycle,* by D. J. Mayhew, 1999, San Francisco: Morgan Kaufmann. Copyright 1999 by D. J. Mayhew. Used with permission.

Portions of this chapter are excerpted or adapted from *Cost-Justifying Usability,* by R. G. Bias and D. J. Mayhew (Eds.), 1994, Boston: Academic press.

Portions of this chapter are excerpted or adapted from "Discount Usability vs. Usability gurus: A Middle ground" by D. J. Mayhew that first appeared on http://taskz.com/ucd_discount_usability_vs_gurus_indepth.htm in September, 2001.

Portions of this chapter also appeared in "Cost Justification" by G. Bias, D. J. Mayhew, and D. Upmanyu, 2003, in J. Jacko and A. Sears (Eds.), *The Handbook of Human–Computer Interaction* (pp. 1202–1212). Mahwah, NJ: Lawrence Erlbaum Associates.

Portions of this chapter also appeared in "Cost Justifying Web Usability" by D. J. Mayhew and R. G. Bias, 2003, in J. Ratner (Ed.), *Human Factors and Web Development* (2nd ed., pp. 63–87). Mahwah, NJ: Lawrence Erlbaum Associates.

REFERENCES

Bias, R. G., & Mayhew, D. J. (Eds.). (1994). *Cost-justifying usability.* Boston: Academic Press.

Bias, R. G., Mayhew, D. J., & Upmanyu, D. (2003). Cost justification. In J. Jacko & A. Sears (Eds), *Handbook of human–computer interaction* (pp. 1202–1212). Mahwah, NJ: Lawrence Erlbaum Associates.

Karat, C. M. (1989). Cost-benefit analysis of osability engineering techniques. *Proceedings of the Human Factors Society 34th Annual Meeting*, 839–843.

Lee, A. Y., & Bowers, A. N. (1997). The effect of multimedia components on learning. *Proceedings of the Human Factors and Ergonomics Society*, 340–344.

Mayhew, D. J., & Bias, R. G. (2003). Cost justifying web usability. In J. Ratner (Ed.), *Human factors and web development* (2nd ed., pp. 63–87). Mahwah, NJ: Lawrence Erlbaum Associates.

Nielsen, J. (Ed.). (1990a). *Designing user interfaces for international use*. Amsterdam: Elsevier.

Nielsen, J. (1993). *Usability engineering*. Morristown, NJ: Academic Press.

Russo, P., & Boor, S. (1993). How fluent is your interface: Designing for international users. *Proceedings of INTERCHI,* 50–51.

Schmitt, E. (1999). *Measuring web success.* Cambridge, MA: Forrester Research.

Sonderegger, P. (1998). *The age of net pragmatism.* Cambridge, MA: Forrester Research.

Sonderegger, P. (2000). *Scenario design.* Cambridge, MA: Forrester Research.

Souza, R. K. (2001). *Get ROI From Design.* Cambridge, MA: Forrester Research.

Whiteside, J., Bennet, J., & Holtzblatt, K. (1988). Usability engineering: Our experience and evolution. In M. Helander (Ed.), *Handbook of human–computer interaction* (pp. 791–817). Amsterdam: North-Holland.

Wixon, D., & Wilson, C. (1997). The usability engineering framework for product design and evaluation. In M. Helander, T. K. Landauer, & P. Prabhu (Eds.), *Handbook of human–computer interaction* (2nd ed.). Englewood Cliffs, NJ: Prentice Hall.

CASE STUDIES

Chapter 9

Cross-Cultural Design for Children in a Cyber Setting

Jorian Clarke
Circle 1 Network, New York and Milwaukee, WI

INTRODUCTION

Different generations grow up with different forms of media for information, communication, and entertainment. The children of today are growing up with new media that allow for real-time, two-way communication of text, graphics, and sound that is simultaneously available around the globe. With the development of the Internet, there is the possibility of a communication channel that can cost-effectively be made available to children from different countries and different socioeconomic levels. Though still not universally accessible, for the first time in history, more of today's children have the opportunity to grow up interacting with peers and with rich information sources using the Internet, providing a global reach and interaction not possible before.

Initially a U.S. phenomenon, the Internet is rapidly expanding into households in other countries at varying adoption levels. Currently, Internet usage figures show the largest number of users is still in the United States, but share of penetration as defined by share of households with Internet access is higher in other countries. As Asia Pacific continues to accelerate in Internet usage, sheer population volume will cause that area to have the largest number of households online in the future. Forecasts indicate a continued trend of adoption in many countries, indicating the need to develop the medium to meet the needs of different users around the world.

Internet usage by children continues to grow due to the fascination this medium holds. This interest holds, for the most part, as children continue online year after year, showing familiarity does not breed boredom. Three key elements that draw children are interactivity, immediacy/availability, and community. In a recent study done on the KidsCom.com site, children predict that the Internet will expand even more into the fabric of everyday usage and most of them believe that their Internet involvement will increase in the next 5 years (SpectraCom, 2001, December)

Children's Internet skills and familiarity are important for all involved in the household. Many families in the United States report that their children are the ones who research information on the Internet for household purchases, thereby contributing to the final purchase by providing information on production/service options, selection, availability, and pricing (SpectraCom, 2001, July). A child's first or deepest exposure to a brand or a character may be online for this generation rather than in a book or magazine, on television or the radio, or in other entertainment venues, such as movie theaters or theme parks. Therefore, it is important that companies provide information in ways that children can understand and utilize.

Finally, children also tend to adopt new technology easier than do adults, as they are not limited by the same concerns that keep their parents from trying new things. Looking at children's current and forecasted Internet usage expectations can help set predictions of technology development in the future, as this generation becomes adults with increased communication needs, skills, and purchasing abilities.

Yet even with these signs of global spread, high user interest, continued usage now and in the future, and the role children play in the household in expanding technology adoption and in impacting decisions with online knowledge, many people who design and develop online marketing entertainment and information are unaware of the issues in designing for children, especially on a global level. Frequently, developers bring biases or use their own childhood experiences in developing for children, instead of researching children on potential content. They also err in using adult logic to set up navigation and graphics, or they do not understand design issues unique to children in age, gender, and global cultural differences.

In addition to understanding user preferences, as the Internet expands from a single point of access to multiple location access (schools, home, Internet cafés, libraries, children's museums, airports, etc.), developers are faced with challenges on how to meet country-specific requirements in this multifaceted delivery channel. That is, how do they find the best design and content development approach for hardware differences, software differences, and access differences?

Another challenge currently more demanding in designing for children than adults in a cyber setting is the inherent issue of privacy and security. How do developers most sensitively and cost-effectively use the full capabilities of this two-way medium to not just push content to children but to publish interactive content, build community interaction based on individual information, and personalize content specific to the child while not violating the potential privacy issues that arise from these opportunities.

First, I examine the interest level of children for this medium. Then, I discuss issues associated with publishing for a cyber setting. Finally, I present the experiences of one of the longest running children's web sites online, www.kidscom.com.

POSITIVE INDICATORS OF CHILDREN'S USAGE OF NEW MEDIA

Global Spread

While there are many sources for counting Internet users and wide ranges from those sources, it is indisputably true to say that Internet usage is spreading around the globe. Twenty nations account for more than 90% of the world's Internet traffic, reports Neilsen/NetRatings service (CyberAtlas, 2002). Currently, the United States has the largest number of users worldwide but share of household penetration is highest in Canada with more than 60% of households with Internet access. Other countries, such as Sweden, Norway, Finland, Germany, New Zealand, Singapore, Ireland, Switzerland, South Korea, and the United Kingdom, also have sizable household penetration (see Table 9.1).

According to a recent report by Jupiter Research, only one quarter of the global Internet population will reside in the United States by 2005, and the Asian Pacific region will contain as much as one third of all Internet consumers worldwide in 2005. The percentage of Internet users in Latin America is expected to increase from 5% in 2000 to 8% in 2005. This growth is attributed to increased PC penetration, telco infrastructure improvements, and reform in those regions. So there are more countries where children will have access to being active online even though not all of these countries will allow children to use the Internet in the same manner (Pastore, 2001).

Children's Current and Future Usage

In addition to expanding the playground exposure to children from other countries, this medium also allows for real-time, two-way communication. As noted in earlier chapters in this book, not all cultures encourage their youth to speak out or to be individuals. However, as a channel for exposure to new ideas from other cultures where children are encouraged to explore

TABLE 9.1

Global Online Populations

			Projection for 2004: 709.1 million (eMarketer) 945 million (Computer Industry Almanac)	
Worldwide Internet Population: 445.9 million (eMarketer) 533 million (Computer Industry Almanac)				
Nation	Population	Internet Users (Source)	Active Users (Nielsen/ NetRatings)	ISPs
Canada	31.6 million	14.2 million (Media Metrix Canada)	8.8 million	760
Finland	5.2 million	2.0 million (eMarketer) 2.15 million (Taloustutkimus)	1.0 million	23
Germany	83 million	26 million (Forsa)	15.1 million	123
Ireland	4.0 million	1.0 million (Amarach)	560,000	22
New Zealand	3.8 million	1.3 million (Nielsen/NetRatings)	1 million	36
Norway	4.5 million	2.2 million (Norsk Gallup)	1.4 million	13
Singapore	4.3 million	1.3 million (Singapore IDA)	956,000	9
South Korea	47.9 million	16.7 million (Gartner Dataquest)	13.1 million	11
Sweden	8.9 million	4.5 million (Nielsen/NetRatings)	3.0 million	29
Switzerland	7.3 million	3.4 million (Nielsen/NetRatings)	1.8 million	44
United Kingdom	59.6 million	33.0 million (Jupiter MMXI)	13.0 million	245
United States	278.0 million	149 million (Computer Industry Almanac)	102.0 million	7,800

Courtesy of CyberAtlas.

and voice opinions, the Internet is stimulating children to have a voice to the world and to each other.

In the longest running study of children and their Internet viewpoints, children revealed their current habits and perceptions of the Internet (SpectraCom, 2001, December). Although this study received participants from many countries, because the study has been fielded in English only up to this time, this leads to sample issues for countries where English is a second language. For those countries, the information is reviewed for findings of a directional but not statistically projectable nature. Half of the children surveyed felt that they use the Internet more now than last year; only 18% said their Internet usage is less this year (see Table 9.2).

In addition, whether they are light users (less than 5 hours per week) or heavy users (more than 10 hours per week), the children predict their 5-year Internet usage to increase to more time (see Table 9.3).

Children's Household Role in New Technology

In the United States, children play an important role in bringing technology utilization into the household. From observational experience, this phenomenon is not exclusive to the United States. In work experiences in Italy, Germany, and France, the younger members of the household ac-

TABLE 9.2
Prior Internet Usage of Children

Last year, do you think you used the Internet:	2001	2002
More time	19%	18%
The same amount of time	31%	33%
Less time	51%	50%

TABLE 9.3
Five-Year Internet Usage Predictions (by Hours per Week)

Five years from now, do you think you will be using the Internet:	Less than 5 hours a week	5 to 10 hours a week	More than 10 hours a week
More time	69%	72%	71%
The same amount of time	19%	19%	20%
Less time	13%	9%	9%

Note. 2002 data only.
Courtesy of SpectraCom Inc.

tively explore new technology and Internet content earlier and with less "technophobia" than adults, often becoming the household "expert" in how to find things online or how technology works. Although not a pattern that holds true in all countries (e.g., children in China are not actively provided content or active online at this time), it is a youth phenomenon that needs to be examined for each country.

In a study fielded globally in English on e-commerce and families done by SpectraCom (2001, July), parents reported that children exercise the same influence and the same decision-making authority online as they do in off-line purchase situations. However, children may be involved earlier on in the purchase cycle. Few children are solo purchasers (i.e., making online purchases by themselves), but they are involved in influencing e-commerce. Fifty-nine percent play the role of requester, asking adults to buy items for them online. Children also act as searchers (looking for products online at a family member's request) or researchers (comparison shopping online at a family member's request). Nearly 25% have either made a solo purchase or made a purchase with the help of a family member (see Table 9.4).

TABLE 9.4
Roles Children Play in Online Purchases

Note. Children can play more than one role, so results total more than 100%. Courtesy of SpectraCom Inc.

Children's Fascination With and Adoption of New Technology

In a recent study by Taylor Nelso Sofres (TNS) the importance of youth in determining the future of the Wireless Internet was highlighted as being one of two key audience targets to drive demand for 3G (Greenspan, 2002).

In the Future of the Internet According to Kids study (SpectraCom, 2001, December), almost two thirds of children showed an interest in owning a mobile phone. In the 2 years this question has been part of this study, there has been an increase in ownership of mobile devices (handheld devices, mobile phones, pocketPCs, pagers) from 32% in 2001 to 41% in 2002 (see Table 9.5).

In order to understand what the development opportunities for alternative Internet devices will be for this age group, the study (SpectraCom, 2001, December) asked them to rate the different devices for a series of different tasks. Their expectation of usage ranged from surfing, communicating via text messages and voice, information delivery, and gaming (see Table 9.6). This will lead to additional design challenges for developers as they strive to reach children's expectations via delivery devices beyond the traditional PC. Due to different adoption rates for different devices already utilized around the globe, this will lead to a variety of devices to design for.

ISSUES SURROUNDING CYBER DESIGN

Delivery Medium Challenge

Designers are challenged in the development of rich media with the multiple access speeds that are available in different countries and usage pene-

TABLE 9.5
Kids and Alternative Internet Devices

Which of these ...	... do you own?		... would you like to own?		... do you think you will get by this summer?
	2001	2002	2001	2002	2002
Handheld device	15%	18%	36%	34%	11%
Mobile phone	22%	28%	60%	63%	14%
PocketPC	5%	7%	51%	39%	8%
Pager	9%	10%	47%	42%	8%
None of the above	68%	59%	11%	12%	70%

Courtesy of SpectraCom Inc.

TABLE 9.6
Five-Year Internet Device Predictions

In five years ...	Handheld device*	Mobile phone/cell phone	Pocket PC/Sony Vaio	Pager
Kids will use that device to surf Web sites.	73%	45%	87%	26%
I will be able to use that device to buy stuff at vending machines.	31%	19%	21%	17%
My friends and I will use that device to send text messages to each other.	89%	84%	87%	70%
My friends and I will use that device to call and talk to each other.	70%	95%	66%	65%
Kids will be able to sign up to have information they want sent to them on that device.	68%	51%	77%	44%
Kids will be able to play games on that device with others.	93%	67%	90%	48%
I will use that device to do my homework.	50%	26%	70%	20%

*a Cybiko, Palm, or Visor.
Courtesy of SpectraCom Inc.

tration within the same country. Narrow band of 14.4K to 56K dial-up modems through phone lines restrict the size of graphics and file sizes for rich media. This restricts content to simple graphics, basic games, and small sound or animation files. In some global locations with limited access, the wait time of 3 to 5 minutes for richer content is simply not worth the cost, especially when children, like adults, find it frustrating to wait for download. Video or extensive sound or animation files that look close to broadcast quality on a broadband connection look choppy and do not stream into a user's machine in a seamless fashion.

Some countries that have led technology may fall behind while countries playing catch-up put in more advanced equipment as they modernize their telecom infrastructures. According to a report from RHK Inc., nearly 50% of the world's DSL subscribers are from Asian Pacific nations (Greenspan, 2001).

This may lead to developers creating content twice: a high-speed multimedia version, for children who have access to broadband and can benefit from richer content delivered through this larger pipeline; and a lower speed version that is limited to static graphics and smaller file sizes for faster download.

Device Evolution

As technology advances, the portfolio of hardware devices will continue to divide into different channels of delivery, including the web (currently the most utilized platform with real-time information delivery, communication, and shopping e-commerce), iDTV (for high-quality audio/visual content delivery), and WAP (for place-based information delivery).

Similar to access options, variety in hardware devices leads to having to create multiple versions of the same content. For example, a game that can be played on your cell phone WAP device will require different graphics than the same game that is delivered to your PC or interactive television Also, a lack of consistent standards across devices means writing multiple versions of code and having multiple servers for delivering content.

Global Challenges in Color, Design Layout, and Language Use

As has been documented in earlier chapters in this book, color associations are learned from the society in which we live. Color symbolism means different things to different cultures and can vary drastically. The situation is even more complex because of gender issues associated with color that are strongly held at younger ages. In different usability studies since 2001 in testing colors for the KidsCom.com web site and for use in designing other sites for children, Circle 1 Network producers found that girls will go into an area that looks like it is for boys (blues, greens, reds), but boys will not go into an area that looks like it is for girls (pink, purple). Background colors and navigation colors are important in helping children know where they are in a site.

Colors for children around the globe are less conservative than for adults. Rich tones and bright colors are responded to better than muted tones and subdued colors (see Table 9.7).

Children's preferences for design layouts also differ from those of adults. Children respond to more stimuli on a page with multiple movement areas, sounds, and more visual navigation clues. They have cognitive literalism, meaning they expect to get what they see when they click on an icon. For example, to find an area on a site with car information, an icon of a car is chosen over an icon of a book showing a car.

TABLE 9.7

Kids' Color Preferences.

What do you think about …	Really cool/ Kind of cool		OK		Not so hot/ Yuk	
	Boys	Girls	Boys	Girls	Boys	Girls
Blue	94%	90%	5%	6%	1%	4%
Purple	42%	86%	20%	9%	38%	5%
Teal	79%	78%	10%	12%	10%	10%
Green	63%	66%	26%	20%	11%	14%
Red	80%	66%	11%	24%	9%	10%
Yellow	35%	44%	45%	37%	21%	20%
Orange	35%	29%	28%	31%	37%	40%
Light Orange	37%	19%	10%	24%	53%	57%

Note. **Bold** indicates statistically significant difference between boys and girls, $\alpha = .05$. Courtesy of SpectraCom Inc.

Small motor skill limitations and limited reading skills at younger ages make mouse click movements easier than keyboard entry or manual dexterity that requires two hands for mouse and keyboard combination entry.

Children use the back button extensively for navigation, and if they are lost in a content search, they have been observed to repeat the same navigation approach, saying "I know this doesn't work," but not knowing what else to try. Sometimes they will then click on anything and in this random approach will "pick up the scent" of what it was they were looking for. For this reason, visual cross promotion of other content combined with clear and simple navigation can aid them in navigating to other areas of interest on a site.

Producers at Circle 1 Network have also observed a difference in countries between the level of slang or popular culture in the use of language that will be tolerated by parents or other adult influencers of children. For example, in translating the English KidsCom.com site into German, producers at Circle 1 Network found from site feedback e-mail that adults were not happy about the informal use of German words in trying to convey a U.S. slang approach.

Language is also best used sparingly in instructions for game play or online creative activity. Girls, who frequently are more mature than boys in their reading and writing skills, enjoy opportunities to practice their skills and will often fully read game or activity instructions. Boys, on the other hand, get easily frustrated if they cannot go right into a game or other activity. They prefer to learn by doing rather than by reading.

We have found visual instructions that are easily accessible by children throughout the activity online are the best approach to supporting their self-help instructional needs. A How to Play icon created for placement throughout the different screens used for an activity is a comfort level to children.

Social Differences and Play Patterns per Gender and Country

An observation by the British Broadcasting Corporation's Theresa Plummer-Andrews in looking at animated programming for the international marketplace pointed out some of the differences off-line that are reflected in online. She mentions idiosyncrasies between countries such as concerns about "emotions" and uneasiness when various animals in an animated TV series were killed off by either man, fire, or old age. She goes on to say that the English find it difficult to tolerate the style of animation produced for the Japanese home market, and the level of violence in some of their series is quite shocking for us, but quite normal for them. However, many hours of Japanese programming finds its way onto French television in its unedited form and no one out there seems to mind it (Plummer-Andrews, 1997).

In observational research, producers at Circle 1 Network we have found that many gender differences cross countries, such as in game play outcomes and in content preferences. In general, boys prefer action games or activities to writing, print-out coloring activities, or emotion-oriented content such as character movies with story lines.

In the development of an action adventure online activity, we learned that both boys and girls had similar interest in "good guy, bad guy" story lines, and got involved in tasks such as collecting "power ups" or "shooting" at obstacles. The biggest differences were that boys did not like having to choose a girl character in their game play whereas girls would use girl or boy characters. In addition, at the end of the game play, the girls wanted to "convert" the bad character to join the group and become a good guy, and the boys just wanted to blow up (destroy) the bad guy!

Communication activities, such as online chat, become of interest to both boys and girls when they are at the age of "cyberdating" or exploring verbal interaction with the opposite sex.

Privacy Issues

Of special note in providing content and marketing products to children online is the aspect of this medium that allows for the collection of personally identifiable information. Special care in designing forms, monitoring children's self-publishing content areas, and warehousing data securely is vital to keeping children safe from outsiders with other purposes in mind. Children have a natural tendency to share information about themselves and their lives. Unfortunately, there is also a conflicting interest by some adults in reaching children for inappropriate contact.

This leads to careful collection and monitoring of areas where children can post information about themselves online that could make them vulnerable off-line. This could be information such as their name, age, school, home address, unsupervised (by adults) play locations and times, siblings' and parents' names, e-mail addresses, and so forth. Site content areas where they may make this information public include registration areas, sweepstakes or contest entry forms, message boards, chat walls, and story or opinion content submission areas.

Each country has different regulations on privacy online and should be researched regularly for legal and regulatory changes. European privacy laws are generally stronger than those existing in North America.

In October 1998, the U.S. Congress enacted the Children's Online Privacy Protection Act (COPPA) requiring commercial web site operators to provide clear notice of their information-gathering practices and obtain prior parental consent when eliciting personal information from children under the age of 13. (More and current information on this can be found at www.ftc.gov.) This law states that any site targeted to children, or any adult site area that has a content area for children or appearing to be for children, needs to get verifiable parental consent before collecting or storing personally identifiable information from children. COPPA also allows parents to access and check the information collected on their children and curtail its use. (Specific details are described on the government web site.)

For developers, this means that a privacy policy needs to be created and displayed wherever content is collected from the user. Companies involved in hosting web site content need to make sure the information is stored and warehoused correctly and access to this information follows the privacy practices indicated in the site's privacy policy.

Privacy regulations per country of service need to be reviewed, implemented, and audited regularly in order to stay compliant.

CASE STUDY: KIDSCOM.COM DEVELOPMENT EXPERIENCES

Site Background and History

KidsCom.com was originally started as a test project in 1994 to see if children were online. The parameters developed for the test were to get 1,000 children to register on the site during 3 months; achievement of these two goals would be considered a strong indicator that children were active online. In 6 weeks, 1,500 children had registered, indicating that the test was even more successful than had been predicted and that many children were actively looking for content for themselves online.

The site at launch was basic with a few simple games, a chat area, an area to search for other children with similar interests, a registration form, and an area to provide feedback to the developers.

Figures 9.1 through 9.4 show some of the design and content iterations the site has gone through over time.

In this initial test, we discovered the significance of this new medium for this generation. Children wanted a two-way communication channel that was available to them at their own schedule. Although the content we created was of interest to them, what caused them to come back repeatedly

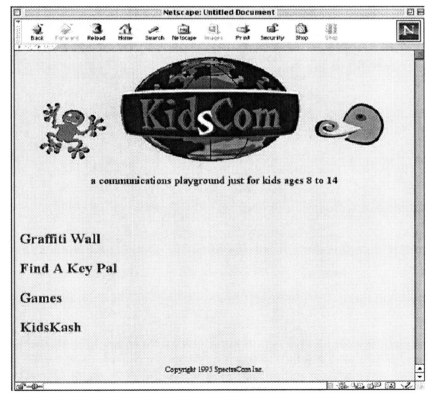

FIG. 9.1. Early 1995 KidsCom.com web site home page. Courtesy of Circle 1 Network Inc.

FIG. 9.2. Subsequent 1996 KidsCom.com web site home page. Courtesy of Circle 1 Network Inc.

was their fascination with the new possibilities of control this technology gave them and the ability to reach out to other children worldwide. We began to experiment with different aspects of this medium and found out that the children were drawn to the Internet even though the ability to provide visuals was not as developed as in broadcast or print. They liked the ability to control their interaction and involvement. This medium gave

FIG. 9.3.   1997 evolution of KidsCom.com web site home page. Courtesy of Circle 1 Network Inc.

them the opportunity to be more than passive observers. With their click choices and keyboard skills, they could determine different outcomes of content and connect in conversations with other children around the globe.

Our first experiences in becoming aware of the importance of this global phenomenon associated with this medium was in the monitoring of the chat wall area and observing the fascination children had with their peers in other countries. Children were amused with different names used to call the same things they shared in their different country environments (e.g., *potato chips* vs. *crisps*, or *lift* vs. *elevator*). The geographical interaction stimulated their sense of place and possibility. They were fascinated to

FIG. 9.4.   2001 evolution of KidsCom.com web site home page. Courtesy of Circle 1 Network Inc.

learn via experiential knowledge sharing of differences and similarities they shared with their peers. For example, while they were in snow and in school, they could talk to other children in other parts of the world where it was summer and school vacation/holiday time. To read about this was one learning experience, but to talk about it with other kids living it opened up a whole new world. This led to the start of our theory that this medium was important due to its interactivity, immediacy/availability, and community.

We also began to hear from parents around the globe about their interest in meeting families worldwide. We also received feedback on content delivery differences based on their equipment limitations and language challenges. At this time, there were not many children's sites in Europe, so many children were using sites created in the United States just to have a place to go online as well as to practice their English.

Global Efforts in Online Publishing

We responded by expanding the publication of the site to include mirrored copies of the content in French, German and Spanish, in addition to the English version. This was our first exposure to the challenges of multilingual publishing. We found that this created much more work than we had anticipated.

We tried a combination of translation by bilingual staff in our central office and a local translation authority for the German portion of the site. The benefit of a local content translator/producer meant that they were familiar with current issues, language idioms, and popular culture. The negative aspects of cross-continent publishing were, at that time, the technology limitations of file sharing and lack of real-time conversation and collaboration of group interaction. The benefit of an onsite translator/producer was the immediacy and group interaction possibility, with the downside being an island of culture apart from the current trends of the country.

We also found customer service in four languages was costly and time consuming. When children feel they have an opportunity to ask questions or submit content of their own creation, they do it with great enthusiasm, more so than adults, and their sense of time is more abbreviated than the time frames adults live in. For example, a child would submit a question through the Contact Us form, and if they did not get a response in a few minutes, they would submit a follow-up question. In the queue of processing e-mail, sometimes a child would get two different responses, the first message processed by one staff member, the second one responded to sometimes by a different staff member. Both employees, not realizing that the person had been sent a response, ended up answering the question in a slightly different way, which would cause the child to send a third e-mail. Many children would get frustrated that there was a wait period of 24 to 72 hours to get a response and send a customer complaint that would then generate additional response time by staff.

In addition, each submission of content created by children needed thorough review prior to posting in order to maintain the standards we had set, which included no foul language, sexual suggestions, or aggressive criticism of another child on the site. This required more labor hours of more expensive staff with bilingual skills to review, respond, and translate.

Other Cyber Setting Issues

The games portions of the site gave us insight into the challenges provided by differing levels of access and equipment quality per country. In the United States, it is fairly easy to find an access provider where costs are a fixed monthly fee for unlimited usage, whereas in many other countries there are telecom costs that limit the number of hours spent online or cause high bills due to per minute usage fees on top of access costs.

This cost per minute of online access in other countries severely limited the amount of time that a child could interact with a game and caused us to consider the development of games that could be downloaded or played off-line.

We learned that game instructions are important to have, but using children's ability to learn from visuals, we have found fewer words and more graphics are vital to building game play understanding (see Figs. 9.5 and 9.6). Children do not like to spend a lot of time reading instructions and often prefer to learn by playing. This behavior can also be observed when children start playing a new video game. The parent frequently learns by reading and wants to understand all icons, rules, and details prior to starting the game. The child will learn by experimental learning done by repetitive game play sessions. This structure of "following the rules" seemed to vary in intensity, influenced by the culture of the country. So the

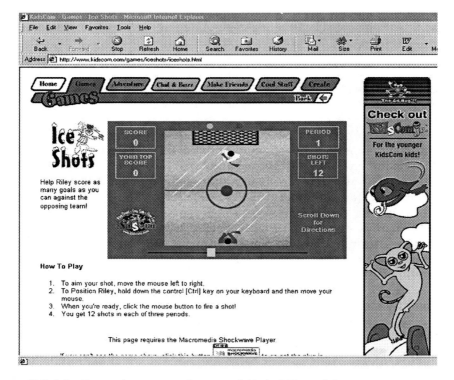

FIG. 9.5.   Instructions as text. Courtesy of Circle 1 Network Inc.

FIG. 9.6.   Instructions as visuals. Courtesy of Circle 1 Network Inc.

development challenge was to keep the game play instruction simple yet have the How to Play button handy throughout the game play in a nonobtrusive, but obvious manner. We are currently in the process of testing instructions delivered in a "play" manner to test for increased comprehension through interactivity.

In addition, through a series of observational research studies in our usability lab, in traditional research facilities, and in home settings, we learned that it is important that a child have success on a game in the first 30 seconds. To continue their play they have to be able to master some part of the game experience in order to stimulate them to learn new things and progress in developing their skill in the game. This success issue in gaming has been consistent across gender, age, and country.

Community Interaction

Access costs, in addition to impacting the way games are developed, can also restrict the amount of community interaction a child in a per minute fee system has with others. Chat walls, where a child may typically spend hours if they have found a community of interest, are not a cost-effective or realistic way for children to satisfy their curiosity about others. So we created our Key

Pal program, an area where kids could search for kids of similar interest in order to start up an e-mail friendship. This form of community building is a more effective way for per minute access fee children to get the community interaction they are looking for. This Key Pal search was conducted online, but once the "match" was identified, e-mail messages could be composed off-line, sent, and downloaded quickly, thereby not having to stay connected online during the time necessary to read and respond with a message. This approach has some limitations for kids, such as it doesn't give the real-time interaction that an online conversation would have. But it has the side benefit of increasing keyboard skills, allowing for deeper and richer interaction because of the stimulus to write, read and edit prior to sending.

Online Research Interaction

It is important when doing any research with children, no matter the country, to get their parents' permission. Children have a natural tendency to want to share information, yet they do not always have the cognitive ability to determine what is appropriate information and what is not. By setting questions up in a nonintrusive manner, and getting parental involvement, much information from children can be discovered in helping to guide content.

Development of surveys for children in a global setting also has specific needs. We have learned to use simple scales with terminology appropriate to children. Depending on the country, informal or slang-type words can be used if they are in the common popular culture (e.g., the word *cool*) in order to engage participation. In countries where children's interaction is more formally guided, strict adherence to correct language usage and structured questions will provide better participation.

To understand differences in illustration style and character preferences, we recently conducted a study with KidsCom children in the United States, Canada, Germany, and Australia. The survey was developed both in English and German. This research sample was further supplemented by recruiting from the German children's site www.kindercampus.de and assisted by a research firm in Germany, Eye-Square, to correctly phrase the study's translation. (See Figs. 9.7 and 9.8.)

The survey is ongoing; however, the sample pulled to date for this publication indicated a difference in children in Germany and the United States in their reaction to characters and illustration styles. German children were more critical of all the characters and more negative in their open-ended responses. More representative human forms and realistic animals were preferred over "fantasy" or "alien" characters by the German children. Whereas U.S. children chose Kimma as their number one favorite, German children commented that "Her head is too big" and "She is too blue" (see Fig. 9.9).

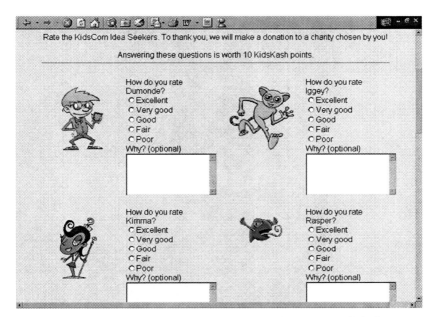

FIG. 9.7. Character research survey page in English. Courtesy of Circle 1 Network Inc.

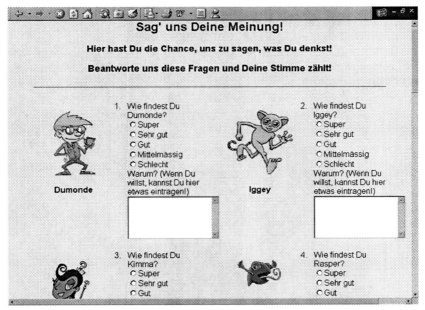

FIG. 9.8. Character research survey page in German. Courtesy of Circle 1 Network Inc.

FIG. 9.9.   Kimma character. Courtesy of Circle 1 Network Inc.

German children's comments about Skeeter included, "He looks cool and like nothing could disturb his peace" and "He looks like he has self-confidence" (see Fig. 9.10).

We are currently continuing to study global differences in children's perceptions of characters used online and how they compare with comic art research available in the field.

Loyalty Program Involvement

Children around the world enjoy the ability to have an off-line reward from an online experience. On KidsCom.com, a loyalty program was developed to encourage children to participate in activities that may be more educationally oriented or timely in nature. Content areas where points are awarded for participation change regularly and include things such as writing areas, geography, and math games.

FIG. 9.10.   Skeeter character. Courtesy of Circle 1 Network Inc.

The area where the program is explained and the reward items are shown is called the Loot Locker (see Fig. 9.11) and children earn points and exchange them in this "virtual world area" for children's products that regularly change. These items include books, fish bowls, food or candy products, video games, and other things of interest to a child. Of note has been the interest in children worldwide in collecting points and "ordering loot." We have shipped items around the world as children send in their "loot orders" for tangible items and experience the fun of the delivery.

Summary & Lessons Learned

The Internet is, and will continue to be, a vital medium in reaching children. There are more new children globally online, and the current child users don't appear to be bored by this communications channel or forecast

FIG. 9.11. The Loot Locker. Courtesy of Circle 1 Network Inc.

decreasing their usage in the next 5 years. Children play a vital role in driving the development of technology and using it in households to guide purchasing decisions.

Even though this age group is active in this medium, many designers are not aware of the design, content development, and other issues that are essential in effectively meeting children's needs. It is therefore important to do research regularly to understand children's preferences, habits, and limitations in different countries and not assume all children everywhere are alike. Access, hardware, software, and alternative device usage are important elements to consider in development. Gender and cultural differences exist in game play, online usage, and color preferences, and privacy regulations are important aspects of a successful online project.

REFERENCES

CyberAtlas Staff. (2002, March 21). *The world's online populations.* Available from http//www.cyberatlas.internet.com/big_picture/geographics/print/0,,5911_151151,00.html

Greenspan, R. (2002, May 24). *Youth and upscale consumers want an m-lifestyle.* Available from http//www.cyberatlas.internet.com/markets/wireless/print/0,,10094_1144611,00.html

Pastore, M. (2001, January 11). *Global Internet population moves away from US.* Available from http//cyberatlas.internet.com/big_picture/geographics/print/0,,5911_558061,00.html

Plummer-Andrews, T. (1997, March 1). *Children's animation, a universal language?* Available from http//mag.awn.com/index.php3?ltype=all&sort=date&article_no=809

SpectraCom. (2001, December). *Future of the Internet according to kids: A longitudinal study.* Milwaukee, WI: Author.

SpectraCom. (2001, July). *The role of children in e-commerce.* Milwaukee, WI: Author.

Chapter 10

Intercultural Human–Machine Systems: Empirical Study of User Requirements in Mainland China

Kerstin Röse
University of Kaiserslautern, Center for Human–Machine Interaction, Kaiserslautern, Germany

INTRODUCTION

The globalization trend of our industrial era has largely intensified cross-cultural communication among different nations. With the increasing exchange of technological products all around the world, the cultural consideration has become an important aspect in human–machine system design. "There is no denying that culture influences human-product interaction" (Hoft, 1996, p. 41). This means that at present when user-oriented machine design is concerned, the cultural background of the target users should be well addressed. In the globalization environment, user-oriented design has also become culture-oriented design. This argument is accepted by both industry and academia.

Germany is one country with obvious export orientation. For German machine producers, cultural consciousness has a great importance for their durable business success in the international market. In 2001, more than 60% of the total German machinery production was oriented to export. The export destinations covered a wide range of international markets, including Western Europe, North America, Asia, Latin America, Eastern Europe, and so on. The non-European market occupied 39.9% of the total German machinery export (VDMA, 2002). This quite various machine export structure motivated a research project at the Center for Human–machine Interaction (ZMMI) in 1996, which aimed to find out the culture-specific user requirements of non-European markets. This project was supported by the German BMBF funds and named INTOPS-1: requirements of the non-European market on machine design. Through an extensive on-the-spot investigation in some Asian and North and South American countries, many special user requirements were revealed. Interesting significant results from this project were published (Zühlke, Romberg, & Röse, 1998).

From 2000 on, based on the observation of the quick economical development in mainland China and its implication on the German machinery industry, one continuous project (INTOPS-2 project) has been carried out in the same center to elicit the culture-specific user requirements of the Chinese market. The previous experience from the INTOPS-1 was taken into consideration in the new project: On the one hand, the previous project revealed that especially in China, the machine users show significantly different requirements on machines, which still need to be further researched (Zühlke, Romberg, & Röse, 1998); on the other hand, the experience from the INTOPS-1 project revealed that the participation of an investigator from the target culture is important to well analyze the information and interpret the investigation data, which has been actually fulfilled in the INTOPS-2 project by incorporating a Chinese researcher in the project group.

In the following sections of this paper, the most important aspects of the INTOPS-2 project, including the approach and interesting or significant results, are presented in the form of an overview.

APPROACH TO INTERCULTURAL HUMAN–MACHINE SYSTEM DESIGN

Culture Research

Cultural influence on human–machine system design can be found in various design aspects. The early research in this area concentrated on the anthropological and physical characteristics of target users from non-Western countries to provide ergonomic design suggestions for technology transfers. In the 1990s, with the quick development of computer science research it focused on the cultural influence on computer user interface design, mainly on the language issues, information presentation format, different information coding, graphical symbols, metaphor, and so forth. Many guidelines were formulated (e.g., del Galdo & Nielsen, 1996; Fernandes, 1995; Russo & Boor, 1993), but the quite invisible cultural influence on human–machine interaction was not identified.

Since mid-1990s, research has concentrated on the quite invisible cultural influences, such as thinking patterns on user interface design issues, menu structure, and information presentation mode (e.g., Choong, 1996; Choong & Salvendy, 1998; Dong & Salvendy, 1999; Shih & Goonetilleke, 1998). Efforts have also been made to deal with the intercultural user interface design in a less pragmatic but more scientific way (Honold, 1999). More and more researchers are trying to connect cross-cultural psychology with user interface design theories to propose theoretical and methodological guidance to culture-specific user interface design (e.g., Plocher, Garg, & Chestnut, 1999).

Up to now, there have been plenty of studies on intercultural human–machine system design, and there have been a lot of applicable results. In the computer software industry, many culture-specific design features have been incorporated in many recent software products for the target markets. However, in the area of production automation industry, the localization of machine user interfaces for the target market is much slower. In the case of machine export to Asian countries, many Western companies have only taken into account basic considerations on user interfaces (e.g., translation for the target market). Other user interface design issues have been neglected. Often, the user interface design standards for Western countries are simply applied to the machines and actually used by other user groups with quite different cultural characteristics. This can result in many problems in human–machine interactions or even cause serious industrial accidents.

As Honold (1999) pointed out, there are many research communities conducting research in the intercultural usability engineering area. Therefore, the results are marked by a wide heterogeneity. No generally accepted design approaches exist, and there is still a large gap between the applicable research results and the practical design requirements. "What is the theoretical foundation that allows one not only to integrate these rather scattered research results, but also to predict where cultural differences will be found in other cultures and products, and how they might be addressed through design?" (Plocher et al., 1999, p. 803). It is clear that a good understanding of cultural aspects and their influence on human–machine system design could help to establish the theoretical foundation in this area.

Essentially, three questions about the intercultural human–machine system design must be answered:

1. Which aspects of culture should be considered?
2. Which machine design issues should be addressed?
3. How can the culture-specific design features be elicited through the analysis of target culture?

Culture Model

Developing a culture model by using some cultural variables is the widely accepted way to study the culture. However, because the existent culture models are mainly targeted at the interpersonal level of cross-cultural communication, their application in intercultural human–machine system design is limited. In his approach of web user interface design, Marcus (2001) used the culture model of Hofstede to connect the five cultural variables (power distance, uncertainty avoidance, individualism vs. collectivism, masculinity vs. femininity, long- vs. short-term orientation) to user interface design issues (metaphor, mental model, navigation, interaction, and presentation; see also Marcus, chap. 3, this volume). This has provided one possibility to connect cultural aspects to user interface design. However, due to the significant differences between the web user interface and the user interface of production systems, the applicability of this culture model in the production system area should be questioned.

A culture model was then developed in the INTOPS-2 project to organize the related cultural data in a way that could be easily applied in machine user interface design. In reference to the culture metamodel of Stewart and Bennett (1991), which classifies culture into two layers—the objective culture and the subjective culture—this INTOPS-culture model classifies the cultural aspects (which should be well addressed in human–machine interaction design) into two groups: cultural mentality and cultural environment (see Fig. 10.1).

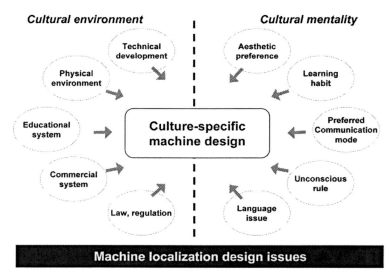

FIG. 10.1.   Cultural mentality and cultural environment.

Cultural mentality refers to the psychological and behavioral characteristics of the user group whereas *cultural environment* refers to the physical and organizational portion of the culture that has significant influence on machine design.

Design Issues

Issues that are affected by cultural conventions should be well addressed in culture-oriented machine design. To date, these issues have been mainly directed toward design issues *within* machine user interface. They include information presentation issues such as color, icon, layout, format, and language. Practical experience has, however, proven that in the area of production automation, especially for machine design for Asian markets, those design issues "*beyond* machine user interface" (Bannon & Bodker, 1991), such as machine functionality, technological features, service organization, are critical for the user's acceptance of machines (Zühlke, Romberg, & Röse, 1998). Therefore, in the INTOPS-2 project, issues not only within user interface, but also beyond user interface, have been considered (see Table 10.1).

Research Approach

One research approach is the description of the way to elicit culture-specific design features. Figure 10.2 shows the research approach of the INTOPS-2 project, which is based on the analysis of Chinese culture. Two phases are characterized in the approach.

In the first phase, the cultural characteristics of the target culture (both in cultural mentality and in cultural environment) are formulated through the analysis of present cultural studies and existent machine design cases. In the second phase, the hypotheses of different design features are generated according to the analyzed cultural characteristics, and these hypothe-

TABLE 10.1
Intercultural Human–Machine System Design Issues

Issues Within User Interfaces	Issues Beyond User Interfaces
Information presentation	Machine functionality
Dialogue and structure	Technological feature
User support	Technical documentation
Interface physical design	Service
Language	General machine design

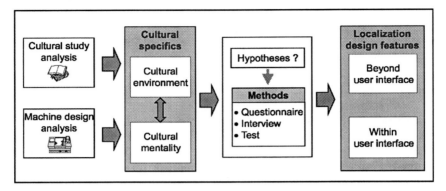

FIG. 10.2.   Approach to culture-specific human–machine system design.

ses are then verified through investigative methods (e.g., test, interview, and questionnaire) to obtain the targeted culture–specific human–machine system design features.

INTOPS-2: INTERCULTURAL HUMAN–MACHINE SYSTEM DESIGN FOR THE CHINESE MARKET

From 2000 on, the project INTOPS-2 has been conducted at the Center for Human–Machine Interaction of the University of Kaiserslautern, in coop-eration with some German machine producers, with the purpose of helping them to localize their machines for the Chinese market more effectively. In order that the final results be used as a practical design aid for the industrial partners in suitably and effectively adapting their machines to the Chinese market, three categories of research topics in the INTOP-2 project were formulated, covering a wide and complete range of localized machine de-sign issues for the Chinese market (see Fig. 10.3).

The approach, which was proposed in the last section, was actually ap-plied in the INTOPS-2 project. In correspondence to the two characterized research phases, the following eight research steps were implemented in the INTOPS-2 project to elicit the machine design features for the Chinese market in practice:

Collection of information material: The information materials regarding Chinese cultural studies and existing machine designs in China were col-lected, which include cultural study literature, Chinese standards, design

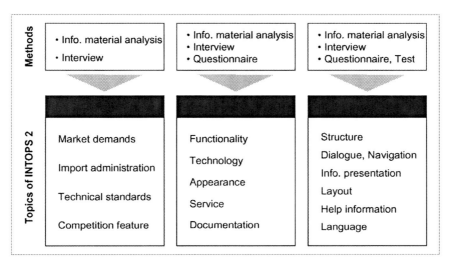

FIG. 10.3. Topics of INTOPS-2 project.

guidelines, machine operation manuals, machine samples, Chinese market research reports, et cetera.

Analysis of information material: according to the related machine design issues for the Chinese market, the information materials were analyzed to get the possible machine design features that are related to the Chinese cultural specifics;

Generation of hypotheses: for those design issues which have at present very few directly applicable results from the cultural studies and existing machine designs, the hypotheses on possible design features for the Chinese market were generated, based on the analysis result of the information material;

Choice of investigation methods: three investigation methods, namely the test, interview, and questionnaire, have been chosen to verify the hypotheses and to elicit the Chinese user's requirements on the machine localization design;

Preparation for the investigation: Cities and organizations to be visited were selected, initial contacts were made, investigation materials (test material, questionnaire, and interview checklist) were prepared and translated, and travel was organized;

Implementation of investigation: the on-the-spot investigation in China was conducted. The realistic requirements of the Chinese machine users were directly investigated through test, interview, and questionnaire. Investigation data were collected in the selected regions in China;

Analysis of the investigation data: the collected investigation data were analyzed through some qualitative and quantitative methods, to verify the previous generated hypotheses and to obtain more meaningful localization design features for the Chinese market;

Interpretation of investigation data: investigation data are further interpreted and summarized to formulate the applicable design suggestions regarding the corresponding machine localization design issues for the Chinese market.

Collection and Analysis of Information Material

Information materials were collected through many possible ways in either Germany or China. The meaningful sorts of information materials include:

Cultural study literature: articles, reports, dissertations, presentations, etc. on cross-cultural communication, methodology and results of culture-specific machine design, Chinese cultural specifics and cognition, Chinese machine user interface features, case studies of machine design practice in China;

Chinese market research reports: statistic reports and government regulations on Chinese economic development trends, technological development of Chinese industry, Chinese machine market demands, import characters and regulations;

Chinese machine design guidelines and standards: Chinese national standards, standardization administration for imported machines, especially the mandatory standards for machine design and Chinese interface design, comparison of Chinese standards to ISO or DIN;

Chinese machine design cases: machine samples and user operation manuals with realistic reflection to general Chinese machine design features and user interface design features (information coding, layout, structure, flow, etc.), typical Chinese documentation for machines, working conditions, working structure.

The information materials were then analyzed to get meaningful information about the Chinese cultural characteristics and the machine design features and interface design features in China. In the analysis phase, all of the collected information materials were structured and classified according to the research topics of INTOPS-2 and their contributions to the project were extracted and analyzed in a comprehensive way. Information materials from different sources were compared to verify the analysis results.

The design features of some superficial culture-specific design issues can be directly obtained from the analysis of the collected information materials. These issues are usually format issues related to information presentation. Other design features could only be obtained through more comprehensive processes, including generating hypotheses and verifying them through methods such as tests, questionnaires, and/or interviews.

Hypotheses on Intercultural Human–Machine System Design for China

Especially for those culture-specific design issues related to the invisible cultural conventions, such as information structure, dialogue design, user support, screen layout, and general machine design features, it is necessary to generate and verify hypotheses. The generation of hypotheses is based on the analysis of the information material. In particular, the following cultural specifics in China have been taken into consideration:

- According to Hofstede's (1997) cultural variables, the Chinese culture is characterized as high uncertainty avoidance, high power distance, obvious long-term orientation, and collectivism.
- The problem-solving strategy of the Chinese can be characterized as systematic and synthetic, wherein understanding of the general problem at the initial stage is very important.
- Chinese culture is polychronic, having the tendency to do many things at once and often changing among them.
- The Chinese language is unique. It is essentially composed of pictorial characters and is more complex in form than Western alphabetic languages.
- Obedience to teachers and elders is much emphasized by the Chinese education system. Students tend to follow teachers' instructions rather than actively solve problems.
- The Chinese education system emphasizes knowledge more than methodology, intensifying the students' tendency to solve a problem using a well-accepted method rather than trial and error.
- Abstract visual art is not popular in China. The Chinese prefer artwork with realistic figures. Traditional Chinese art is characterized by symmetry.
- Technical standards in China still differ in many ways from international standards, including the application of colors, symbols, and so on.

- Interpersonal communication between the Chinese is considered high-context communication, with most meaning contained within the context. What is said is often not as important as what is not said.

Based on these cultural specifics in China, several hypotheses categorized here in 7 groups were generated. A brief list of these hypotheses is listed below:

Hypotheses on general machine design

Chinese machine users prefer familiar general machine design features (in form, color) and would reject design features that are foreign to them; that is, they accept machines with mature and familiar technologies rather than totally new technologies with which they are not familiar;

Hypotheses on general interaction concept

Chinese machine operators prefer a standard configuration of machine user interface with low flexibility for configuring on their own;

Hypotheses on menu structure

The Chinese machine user interface has a wider but less deep menu structure with more menu items appearing on the high (surface) level of the structure;

The menu structure of the Chinese machine user interface should facilitate the parallel processing of more than one task more conveniently;

Hypotheses on interaction navigation

The navigation of the Chinese user interface should provide very strict and definite ways to follow rather than to allow the operators to decide how to proceed further;

Hypotheses on user interface layout

All the menu items of the main menu and second menu layer should better be laid out on the first display mask of the Chinese machine user interface;

The user interface should provide visualization methods to show the general menu structure on display;

Chinese machine operators prefer the vertical menu layout of the Chinese interface to the horizontal one;

Chinese users prefer the symmetrical layout of information on a user interface;

Hypotheses on information coding

The color coding of the Chinese machine user interface is different from that of the Western standard. Western color coding application can not be correctly interpreted by the Chinese machine operators;

The use of the Chinese textual characters, together with graphical symbols, is typical of the Chinese machine user interface;

Hypotheses on user support information

Support information is required by Chinese operators for basic and simple machine functions. For complicated problems, they tend to search for help from colleagues instead of from the machine user interface;

Help information is better organized in the form of a tutorial program to facilitate the learning of the operator in this process;

Graphical presentation of the help information is dominant over textual presentation in presenting the operation process;

Help information concentrates more on the presentation of the operation process than on the explanation of technical principles of the machine.

Investigation Methods

The following three investigation methods are applied in the INTOPS-2 project: test, questionnaire, and interview. These methods have been proven by the previous project INTOPS-1 to be quite effective in investigating user requirements on culture-specific machine design (see also Zühlke, Romberg, & Röse, 1998). However, due to the different purpose of these two projects (demonstrating cultural influence on machine design for INTOPS-1 and finding out definite design features on the Chinese market for INTOPS-2), wider and more detailed requirements of the Chinese machine users on machine design have been concerned. Therefore, a few new tests and more detailed questionnaire and interview checklist have been developed for the INTOPS-2 project. An overview of all the implemented tests is presented in Table 10.2. The main information categories within the questionnaire and the interview checklist are shown in Table 10.3

Furthermore, another quite critical change for the INTOPS-2 project is the incorporation of a native-speaking Chinese researcher in the project group. This has proven to be an indispensable measure, ensuring the effectiveness of the on-the-spot investigation.

The three investigation methods in this project complement each other. The tests are mainly formulated to verify the hypotheses concerning user interface design. The questionnaire is mainly applied to elicit user require-

TABLE 10.2

Overview of Tests Used in INTOPS-2

No.	Test	Aim	Material	Subject	Analysis
1	Preference to color composition for machine tools	Elicit preferred color composition and difference to German one	10 cards with differently colored machine tools	No special requirement	Average preference degree for each composition
2	Recall performance for graphical information vs. Textual information	Test information processing ability for different information presentation modes	3 pieces of paper with different modes of presentation: • text only • picture only • text & picture	No special requirement	Average recall rate for each mode; Characters for better information recall
3	Understanding of color coding	Test understanding of standard color coding and difference to German one	7 standard colors of IEC; 3 groups of concepts in daily life and at work (5 in each one)	Matching for concepts at work only for machine operators	The color association rate for each concept
4	Symbol understanding	Test understanding of standard ISO symbols and elicit the preferred symbol characteristics for information coding	Icons from ISO and Windows; 2 kinds of materials: 18 icons, each with 3 possible meanings; 14 meanings, each with 3 possible icons	Machine operators	Average recognition rate for each icon; Character for better matched icon

5	Familiarity with Windows interface	Test familiarity with the Windows interface	Integrated with Test 4	Machine operators	Recognition rate for Windows icons
6	Concept of grouping	Elicit the grouping rule and the difference to German one	74 cards with different CNC (computer numeric-controlled) machine functions	Only with experienced CNC machine operators	Preferred structure for grouping
7	Preference to screen layout	Elicit familiar screen layout characters and difference to German one	Over 20 different cards in form and size representing the screen elements	CNC machine operators	Preferred layout of different screen elements
8	Understanding of English terms	Test English comprehension	One table with 54 English technical terms	Machine operators	Average understanding rate; Character for better understanding

TABLE 10.3

Categories of the Questionnaire and Interview Checklist

Questionnaire	Interview Checklist
Basic information about the visited company	Basic information about the visited company
Information about machine purchasing	Information about machine purchasing
Requirements on machine service	Application situation of imported machines in practice
Requirements on machine user interface	Requirements on machine user interface
Requirements on machine technical documentation	Requirements on technical documentation
	Requirements on service
	Information about work organization and training

ments on general machine design features and functional features. These two methods are both highly structured and can be well controlled by the investigator in practice, but at the same time they have the disadvantage of restricting the investigation within the definite areas that have been formulated by the investigators. Therefore, they are not capable of eliciting those design features that are not imagined by the investigator. In comparison, the interview could, in turn, provide the interviewees with more freedom to express their requirements in wide range, which could help the investigator to obtain knowledge of wider and unexpected requirements of the target market. The combination of these three investigation methods was successful and effective. The application of these investigation methods in INTOPS-2 to elicit of design features for different design issues is listed in Table 10.4.

Before the on-the-spot investigation, different investigation materials were prepared in Germany, including the test materials for a total of eight tests (see Table 10.2), one questionnaire, and one interview checklist. All of these materials were translated into Chinese.

On-the-Spot Investigation

The on-the-spot investigation in China was carried out at the end of 2000 until early 2001. During three months, 32 Chinese organizations in Shanghai, Beijing, and Chongqing were visited, of which 26 were Chinese industrial enterprises (including Chinese machine tool producers and

TABLE 10.4

**Application of Different Investigation Methods
for Different Design Issues**

Design Issues		Investigation Methods		
		Interview	*Questionnaire*	*Test*
Beyond UI	Machine functionality	++	++	
	Technological features	++	++	
	Technical documentation	+	++	+
	Service	++	+	
	General machine design features	+	++	+
Without UI	Information presentation	+	+	++
	Dialog and structure	++	+	++
	User support	+	+	+
	Interface physical design	+	+	
	Language issues	+	+	+

+ Indicates that these methods are well suitable.

some machine users). The other 6 organizations included governmental organizations for import administration and research institutes for machine user interface design in China.

Even though the investigation material had been carefully checked in Germany before the investigation, in practice, a few investigation materials caused problems due to the format or expression of some terms. One example was with the printing of symbols on separate pieces of paper in the recall test, causing subjects to turn pages to remember the symbols. This inconvenience had not been found in the pretest phase because at that time, the test material had not yet been bound. Following discovery of this problem, the test was printed with all symbols appearing on one piece of paper. Other small adjustments were made to the questionnaire and interview checklist.

The *tests* were usually conducted in the break room. During the test process, the presence of other persons in the test area was, if possible, avoided. The subjects were all Chinese machine operators. A total of 42 subjects participated in the tests, most of them in more than one test. The total test span was controlled within 30 minutes, which has proven in practice as almost the maximum acceptable time for a subject to willingly participate. Due to this constraint, none of the subjects participated in more than four tests.

The *interviews* were mainly conducted in the office of the interviewees. Sometimes also other persons were present with the interviewees to accompany the investigator. Sometimes more than one interviewee with different responsibilities in one firm were separately involved. Since the investigator was a native speaker of Chinese, there was no problem in language communication. All interviews were conducted in Chinese.

A total of 35 main interviewees were involved. Half of the interviews were conducted in Shanghai and the others in Beijing and Chongqing; 58% of the interviewees were machine tool users in China (in automobile, motorcycle, and machine tool industry sectors). The interviewees were either responsible for the machine purchasing decision (the chief manager, chief engineer, and factory/workshop director) or responsible for the actual machine application in practice (the chief engineer, factory/workshop director, equipment engineer ,and technician).

Most of the *questionnaires* were filled out by the same interviewees directly after the interviews or by other responsible persons in the firms. A few questionnaires were left to interviewees to fill out afterward and then were sent to the investigator via fax. In comparison, the former way is much better than the latter because any lack of clarity in the questions could directly be explained by the investigator. A total of 19 questionnaires were obtained in the investigation. More than half of the questionnaires came from Shanghai. About 63% questionnaires were filled out by the state-owned firms in China; 47% of the questionnaires were filled out by the machine tool users in China (in automobile, motorcycle, and machine tool industry sectors).

SELECTED RESULTS OF THE INVESTIGATION IN CHINA

Plenty of investigation data were obtained in China. The analysis of these data was conducted using both qualitative and quantitative methods. Some really important intercultural human–machine design features were revealed by the investigation. The interpretation of these investigation results would lead to several important localization design features of use to German machine producers adapting their machines to the requirements of the Chinese market.

Requirements on Design Issues Beyond User Interface

As was mentioned, some user requirements on design issues beyond user interface were elicited through the use of questionnaire and interview. In the project INTOPS-2, the user requirements on these issues are concentrated in the following aspects:

Quality and Technology. High machine quality and advanced technology in imported machines are still critical characteristics for attracting Chinese machine users. In many cases, these characteristics are the most important decisive factor influencing the machine purchase. Very often, the requirements on machine quality and technology are largely influenced by Chinese import policies or other administrative regulations.

In comparison to the domestic machines in China, the imported machines have, at present, a technological advantage, especially in the following areas, which could facilitate their competition capability within the Chinese market:

- The ability to provide machines with significantly higher working accuracy, higher running reliability, wider working range, and higher productivity than the other competitors.
- The technical collaboration ability to provide the complete set of production equipment for large-scale construction projects.
- The application of information technology in the machines to reinforce their communication between each other and to connect to local area networks of users.

It should be emphasized that the quality and technology advantage of imported machines might be largely influenced by environmental factors in China. Therefore, special measures should be taken to ensure that the machine can also reach the expected technological specifications, even under the following specific physical environment conditions, which are characteristic of Chinese factories: (a) high humidity at the workplace, (b) high concentration of acid substance in the air, (c) high concentration of particulates in the air, (d) insufficient air conditioning, (e) large temperature differences between different seasons, (f) high fluctuation in voltage and frequency in electricity supply, and (g) high frequency of electricity blackout.

Functionality.   User type, scale, product range, work organization, and personnel qualification vary greatly in China. Requirements on machine functions are also very different. However, there are still two common requirements that are characteristic of Chinese machine users. First, the requirements on machine functionalities are pragmatic. In most cases, only those machine functions applicable to present production needs are required. Second, Chinese machine users emphasize stability and continuity, which implies that they would not change the machine too often but would upgrade it when necessary.

The all-in-one machine design concept will make the machines too complex and too expensive. The specialized design of machine functions will cause a too wide product spectrum and leave little room for functionality promotion; therefore, neither is suitable for the Chinese market. The best way is the modularization design. This way could provide user requirements on flexibility, to configure machine functionality according to their pragmatic needs, and the possibility to promote machine functions according to their future application changes.

Furthermore, considering the high requirement on machine working quality and the low qualification of Chinese machine operators for effective machine control, the Chinese machine users require that the machine control functions should be accomplished by automatic systems without much interference of the operator. This implies that the machine control functions that are critical to the final machine working quality comprise the basis of the machine modularization design. On the other hand, because of the inexpensive work force in China, many Chinese machine users require that many material handling functions (which could be well accomplished through manual work) should be allocated to their workers instead of to the automatic machine system. These functions, mainly the auxiliary functions in material handling, comprise the most significant part of the functions to be modularized.

Appearance Features. The imported machines have one special symbolic meaning for the Chinese machine users and their customers. Therefore they have mainly the requirements on machine appearance features that they should be easily recognized as imported machines through differentiating appearance to domestic machines. The visual appearance features such as form, color composition, machine brand name visualization, etc. should have special characteristics. In some cases, the consideration on these aspects could to some extent influence the user's decision to purchase imported machines. The collected information material from China shows that the Chinese machines have their own appearance features. For example, the colors blue and green are dominant but red and purple are seldom, the brand names have ascertain and the same with the model numbers. (e.g., 914: they will die, 014: you will die, 740 have bad associations in Chinese).

In most of the cases, the requirement on machine appearance feature is not so important as that on machine quality, service and price, etc. and is not a decisive factor in machine purchasing. The translation of machine brand name in Chinese, the sound of the machine model names, etc. should be well considered to reinforce the positive feeling of the machine and avoid some negative associations. Especially the application of some unfortunate names in China should be avoided.

Technical Documentation. Technical documents of imported machines are a basic and important support for the Chinese users to effectively apply the machine. However, the actual situation of technical documentation of imported machines in China has not met the user's requirements in many aspects. Very often the machine producers provide only the original technical documents (usually in English) for their Chinese customers. For those who have already provided the Chinese technical documents, the translation quality is very often unsatisfactory.

The users required that the machine producers should well organize the localization of technical documents into one important aspect of their product development. Even in the least scale, the most basic requirement of the target users on the technical documentation, e.g. language, should be well fulfilled.

Document Language. The language requirements are the most basic requirements for the technical documentations. Depending on different machines and different machine users, the language requirements of technical documentations can be fulfilled in different scales. But no matter how, the following aspects regarding language should be well considered:

- In principle, all the descriptions except some internationally acknowledged technical terms and abbreviations on technical documents should be translated into Chinese. However, because of the high translation cost, sometimes only a part of the whole technical documents is translated. In this case, it should be ensured that the machine operation part should be translated at first. This requirement is a little different from that of machine user interface, because the operation manual involves mainly the textual description which can not be well understood by the operators;
- Especially important for the foreign machine producers, is that there are significant differences in language habits of different Chinese spoken areas, such as Hong Kong, Taiwan, Singapore, and Mainland China which would largely influence the understanding of the translation;
- No matter who has accomplished the translation work, the translation should be checked in practical application for some possible misunderstanding of the operators. The machine producers should also check the usability of their technical documents in practice and then to revise them if necessary;
- The language character of textual description of the operation manual and maintenance instruction should well correspond with understanding ability of the machine operators and maintenance technicians. For the operation manual the language should be more clear and short. The description of some important operations should be supplied with a brief description of the operation effect. For the maintenance instruction the description should be more detailed and complete. The description of some technical principles should be supplied.

Document Form and Content. The document characteristics on its type, content, structure, format, and so on, should be well addressed according to the target user's learning habit and working organization. Usually, the machine producers should still do some further investigation regarding these detailed requirements on technical documents for definite types of machines on the Chinese market. Some general suggestions for well preparing the technical documents for use in China are summarized below:

- A complete set of technical documentation should be supplied to the Chinese machine users. For the purpose of maintaining machines by themselves, many users in China emphasize the need of some technical drawings, including electrical connection plan and mechanical outfit drawing of some machine parts;
- A good overview of all the provided technical documents should be prepared for good document management. Different documents should be well labeled to have good references for the Chinese users. It was found in the investigation that many Chinese users don't even know which documents should have been provided for their machines and they don't know where these documents are;
- For the sake of good documentation management, measures should be taken to provide documents which could be more easily preserved by the Chinese users. The traditional paper document form is proven to be easily lost in China because of ineffective management organization. Other alternatives, especially the electrical forms of documents, or some back-up forms of documents for example on the internet, should be considered to supply in China in the future;
- Considering that the Chinese machine operators are always trained on the job at the machines, it is suggested that the operation manual should be supplied in combination with the machine help information on the user interface to facilitate the operators learning machine operations more conveniently during their work;
- Realistic pictorial illustrations to describe the machine operation and maintenance procedures are much preferred by the Chinese users. But the pictorial presentations should be well arranged with the textual descriptions in the documents. At the same time the format and layout information to differentiate various kinds of operations and their structure should be well utilized.

Machine Service.   Because of the general lack of high qualified maintenance technicians in many Chinese companies and the critical importance of imported machines to their production (the high investigation correlates with very high exploitation), the Chinese machine users have very high requirements on service for imported machines. The service quality of the foreign machine producers influences on a large scale the final machine usability in the user's factory.

In China the present service quality of many foreign machine producers is very problematic. Many services provided in China can not compared to the services the producers provided elsewhere. The user's interest in service is often neglected. However, with the increase of more foreign machines on the Chinese market, the present lack of service could be fatal to their further business in China for some foreign machine producers. The following requirements on machine service are characterized in China:

- With the presentation of machines on the Chinese market the foreign machine producers should establish a good local service department which covers the service in before-sale, sale, and after-sale loops. Especially the before-sale service for machine choice consultation and the after-sale service for machine problem solving is a lack in China;
- The most important requirements of Chinese users are after-sale service including the quick and effective response to service requests and prompt spare parts supply. Most Chinese users emphasized that there should be one service center in Mainland China (not in Hong Kong or Taiwan) with qualified service engineers. The best way for spare parts supply would be a regular spare part bank in Mainland China. If this is impossible, other measures in machine design, in service organization etc. should be taken to reduce the delivery time effectively;
- Considering the geographical constrain to effective machine service, some tele-service alternatives such as through telephone, internet, etc. are practical and should be used much more by the foreign machine producers in the future;
- Training for machine operators as well as for maintenance technicians is much required by the Chinese machine users, especially the effective training for maintenance technicians is of high importance for them. More complete training programs should be provided in China, including some regular or irregular training for machine operators or maintenance technicians after some time of machine application.

Requirements on Design Issues Within User Interface

The user's requirements on human–machine system issues within user interface have been elicited through either tests, or interviews and questionnaires. The investigation results have indicated that although the present competitive aspects of imported machines on the Chinese market are still mainly in machine quality, price, service, etc., in the near future, the competition in user-oriented machine design and user friendliness will be a significant factor to distinguish different machine producers. Even today, a special approach for user interface aspects could also add extra machine competition capability on the market. The main requirements on the machine user interface design are summarized next.

Language.   The interaction language requirements are the most basic and important requirements of the Chinese users for machine user interface. For most of the Chinese users the machine user interface in Chinese is a significant sign that the machines have been localized in China. A Chinese user interface is usually taken as the largest convenience provided by the foreign machine producers, because it could give the Chinese machine users a direct confidence that the foreign machine could be well operated without great understanding problems.

The test of English term comprehension revealed that the Chinese machine operators have generally quite limited understanding ability for English terms applied in machine operations. The understanding rate for all tested English terms is 46.6%. The correct understanding rate for some representative English terms is shown in Table 10.5. The best understood English terms include: *OK* (95.7%), *OFF* (82.6%), *Stop* (82.6%), *Home* (73.9%), *Help* (73.9%), *kg* (73.9%), *ON* (69.6%), *Start* (69.6%), *mm* (69.6%), etc. They are all terms which are related to the basic and most often met machine operations (on/off, start/stop), or the terms which could be often met in daily life (ok, home, help, kg). The worst understood English terms include: *PIN* (8.7%), *Icon* (13.0%), *Esc* (21.7%), *Undo* (21.7%), *Disk* (21.7%), *inch* (26.1%), *rpm* (26.1%), *Edit* (30.4%), *Menu* (39.1), *Shift* (39.1%), etc.. Most of these terms are related to the computerized system operations (PIN, icon, disk, edit, menu, shift), which implies that most of the Chinese machine operators are still not familiar with the operation of the computerized control systems. The test results have demonstrated the importance of a Chinese user interface version for the effective and safe machine operation in China.

However, the present language issues of the foreign machine user interface in China are not satisfactory. Only a few of them are provided with Chinese user interfaces. A large part of other machines are provided only with user interfaces in a foreign language, usually in English (sometimes also in German, in Japanese, etc.).

TABLE 10.5
Comprehension Rate of Selected English Terms

Terms	OK	OFF	Stop	Home	Help	kg	ON	Start	mm	End	Enter	Del
%	95.7	82.6	82.6	73.9	73.9	73.9	69.6	69.6	69.6	65.2	56.5	52.2
Terms	Exit	Reset	Shift	Menu	Edit	rpm	inch	Disk	Undo	Esc	Icon	PIN
%	43.5	39.1	39.1	39.1	30.4	26.1	26.1	26.1	21.7	21.7	13.0	8.7

It is required by the Chinese machine users that the language issues should be firstly addressed for the machine with the aim to be used in China. However, because the user's requirement on this issue is different, in practice the foreign machine producers have some flexibility to handle this localization issue according to their localization capability. Some suggestions for them to well treat this aspect could be:

- Localization of machine user interface regarding language can be divided into different levels. Two levels of localization are significantly differentiated: the translation of machine labels into Chinese and the translating of the screen display into Chinese. The former concerns only the reprinting of indications and text descriptions on the machine surface into Chinese but it could give the first and direct impression of localization design. The later concerns the reediting of the software codes is more complicated and expensive than the former one;
- The translation of screen displays into Chinese concerns integrating one Chinese character databank into the control software and reediting those software parts which present screen information. It includes mainly the work of replacing the English text or words with Chinese or corresponding characters. This involves two levels of work. The first level is to translate only those parts directly related to machine operations, which are mainly some indications or simple text description on the screen. The second level is to translate all the screen displays, including the information for machine maintenance into Chinese. Usually only the first level is fulfilled, because the amount of work is much less that the complete translation. The investigation results, however, have indicated that for the Chinese machine users the translation of the machine maintenance part is more important than that of the operation part. With this part there are more understanding problems. This requirement from the investigation is quite different to the current localization cases;
- In many cases, the machine user interface would be used by different persons with different language backgrounds, for example, the service engineers from Germany teach the machine operation in China. Therefore, if it is possible, one dual language user interface would be a good alternative to the user interface separated for each operator or the preselection of language and in dependence of the individual user profile;
- It is clear from the investigation that the requirement on interaction language is at present only on the information presentation, namely the display aspect. There are some requirements on infor-

mation input aspects, because the Chinese operators don't need to input the Chinese characters. Most of the input (e.g., programming) could be accomplished through some codes. This means that the Chinese input is not yet needed for present machines and the input could still be kept as alphabetic.

Menu Structure.   It was assumed before the investigation that the different requirements of the Chinese users on the menu structure would be influenced by the different thinking patterns of the Chinese to the German. However, in practice, the most significantly (or most obviously) different requirement on this aspect comes from the different working organization. One of the significant characters of working organization in China is the more elaborate division of operation tasks. A Chinese operator's task range is more narrow than that for a German machine operator. The effectiveness of the German machine user interface to support the working organization in China is problematic. In fact, this problem has been reflected by some Chinese interviewees they require a self-defined user interface to fit their individual production tasks. Some requirements regarding the menu structure are proposed by the Chinese machine users:

- At first, the menu structure should well correspond to the operation tasks of the Chinese machine operators. The new menu structure should be characterized by the separated (and hierarchical) visiting right of different machine operators to different machine functions, with a simpler menu structure for each operator group. Especially for the machine operators the operation functions should be much simpler. This could also make the operation of definite machine function more quickly to be reached;
- Because the actual working organization for each customer is different, it is impossible for the machine producers to provide for each customer his one individually menu structure. In practice, there should be a balance of consistence in menu structure provided by the machine producers and the flexibility of users to configure the structure especially for his definite working organization. Based on this consideration, the modularization of the menu structure to leave room for the further adaptation is a good alternative. Then the menu structure for a specific customer could be configured by the producer according to his individual need. This freedom to define their own user interface through some openness of the system has been required by many Chinese machine users.

Dialogue and Navigation.   Some direct or indirect information has been obtained from the investigation which is involved in the design of dialogue forms and navigation system for the user interface. From the problems that the interviewees often reflected about their operators and the self-evaluation of the machine operators about their own work, some requirements of the Chinese users on machine dialogue and navigation system is revealed. The following points should be helpful for the foreign producers to design user interface for the Chinese operators:

- The Chinese customers have very low assessment of their own machine operators. The operators themselves have also very low evaluation about themselves. It is often remarked by both of them that the operators have quite low qualification and could not understand the complicated machine operations. This reflects that the machine operators could have potential fear to actively interact with the machine and would prefer to follow the definite operation instructions from the system. It is then required that the dialogue system should provide more error tolerance to the operation and should have a very clear guide for the operators to let them follow the operation process;
- Secondly, from the questionnaire results it can be found that the main problem of the Chinese machine operators in machine operation are based in the understanding of different machine functions and different operation processes. It is then required that the user interface should provide a navigation concept which presents the general machine functions and operation processes in an overview way, to let the operators establish one clearer overall image of machine operation more easily;
- Because of the general doubt of the Chinese machine users about the qualification of their operators, it is required that the possibility to intervene by operators in machine control operation should be low. This means that the machine options and settings should be more standardized and offer minimal possibilities for the operators to change. It is then suggested that those dialogue forms should chosen which have only few options.

Interface Layout. It was hypothesized that the cultural influences from language (the reading direction) and aesthetic preference should have large influence on preference to screen layout. However, the screen layout test has indicated, the preference of the Chinese machine operators to machine user interface layout is largely influenced by their actual working experience with some existent machine user interfaces. This result also means that for the design of machine user interface layout for the Chinese market, the existent user interfaces should be well investigated. Especially the following two points should be considered in interface layout design:

- The working experience of the Chinese machine operator with the existent user interface has significant influence on their preference to interface layout. This means that principally all kinds of interface layouts could be well adapted and accepted by the Chinese machine operators in the long run. It is then suggested that the widely applied user interface layouts in the Western countries could be applied in China without significant changes. This superficially "simple" way could even be a better way to facilitate Chinese operators to adapt themselves more quickly to the imported machines in the future;
- The Chinese machine operators have little experience with the operation of Windows interfaces. Therefore the application of the Windows concept for machine operation in China will meet some problems at the beginning. Furthermore, the quite free dialog provided by Windows interaction could also make the Chinese machine operators more unsure about their operations than the guided dialog. It is then suggested that the application of the Windows interface in China is not encouraged for machine operation at present.

Help Information. Chinese machine operators rely much on help to operate the machines. At present, this help is mainly from their colleagues or masters. The machine system, including the technical documents, have provided only little help for the Chinese machine operators. In the investigation, many Chinese machine operators expressed their requirements for more help on the machine user interface, which includes not only the machine trouble diagnosis help information, but also the help for normal machine operation or learning. The help information on machine user interfaces is, in some ways, even more important than the accompanied machine operation manuals, because the operators are always trained on the job at the machine, and the operation manuals are not allocated to every operator. It is then required that the provision of more effective help information on user interface should be completed by the machine producers.

Information Coding. Although different information coding ways can be found on machine user interface in China, the most preferred way is through Chinese *text*. Almost all Chinese domestic machines are labeled with Chinese text to indicate keys, buttons, and other information. The foreign machines which have been provided no Chinese labels have also been very often labeled by the Chinese users themselves with Chinese text. The interview results have also revealed that the Chinese machine operators prefer the Chinese text label, when they are required to choose among different information coding ways (e.g., color, symbol). For them, only the text label can express the operation information most clearly without any misunderstanding. The Chinese text is the most reliable and explicit information coding in China.

The results of the English comprehension test revealed that the Chinese machine operators have generally unsatisfactory English comprehension of safe and effective machine operation. Therefore, the text coding in English should be restricted to a very limited scope. More detailed requirements at this point are summarized:

- The most important information should be at any time encoded and usable Chinese text (and with combination to other information coding ways such as color, shape and position, etc.);
- Only when it is really necessary, the English texts which express some general machine operations such as On, Off, Start, Stop, etc. and are closely related to some daily use such as Ok, Yes, Help, Home, etc. could be applied for information coding. But the application of most English abbreviation in text labeling should be avoided;
- In many cases, other information coding ways should be applied together with the Chinese text to ensure that the machine can be well maintained by foreign service engineers.

The results of the test of understanding of *color* coding are presented in Table 10.6.

Contrary to the previous hypothesis, the results have revealed that the Chinese machine operators could already well interpret the international standard color coding in working environment (See IEC73). Especially the application of color *red* (danger), *green* (normal), and *yellow* (caution) to express meanings in working context could not meet any serious problems. This implies that the standard color coding applied in Western countries could also be applied in China.

The test results have also revealed that the working experience has large influence on the understanding of color coding in China. From Table 10.6 it can be seen that the highest color association in daily life is 70,4% (red with luck) and the lowest is 25,9% (blue and black with power), while the highest color association in working context is 92,6% (red with danger) and the lowest is 40,7% (cyan with OK).

Based on the basic test results, regarding the color application on user interface design two points should be well considered:

- The international standard color coding is suggested to be consistently applied on machine user interface for the Chinese market. For the Chinese machine operators the working experience with machine operation is the main influence on the interpretation of color coding. Therefore the consistent application of an international color coding standard could intensify their understanding;

TABLE 10.6
Percentage Association of Each Color With Each Concept

Concept		Color							Total (%)
		Black	Green	Cyan	Blue	Magenta	Red	Yellow	
Daily life	Luck	—	7.4	7.4	3.7	7.4	70.4	3.7	100
	Fear	59.3	11.1	3.7	14.8	3.7	—	7.4	100
	Hope	—	40.7	11.1	3.7	7.4	7.4	29.6	100
	Silence	3.7	14.8	55.6	7.4	3.7	—	14.8	100
	Power	25.9	7.4	3.7	25.9	11.1	14.8	11.1	100
Information class	Danger	—	—	—	—	—	92.6	7.4	100
	Caution	3.7	3.7	3.7	—	11.1	3.7	74.1	100
	Normal	—	74.1	11.1	7.4	3.7	—	3.7	100
	Regulation	44.4	—	22.2	25.9	3.7	—	3.7	100
	General Info	11.1	14.8	44.4	18.5	3.7	—	7.4	100
Operation	On	7.4	74.1	3.7	—	—	7.4	7.4	100
	Off	7.4	3.7	—	—	—	85.2	3.7	100
	Help	—	3.7	14.8	3.7	11.1	—	66.7	100
	OK	7.4	7.4	40.7	18.5	14.8	3.7	7.4	100
	Exit	44.4	7.4	18.5	3.7	18.5	—	7.4	100
Total association		58	73	65	36	27	77	69	

- The application of color coding should be limited within possibly fewer colors in order to transfer the information more clearly. The colors of *red*, *green*, and *yellow* are suggested to be applied whenever it is necessary. The application of other colors for information coding should be more careful and restricted within some uncritical information classes. Usually other information coding methods should be applied together to ensure correct interpretation.

Although *symbols* and *icons* are important information coding ways to cross over language barrier among different countries, the general understanding ability of the Chinese machine operators to different international standard symbols and icons is very low. The average understanding rate for all the tested icons and symbols is 44%; of which the average recognition rate for

ISO symbols is 45,7% and that for Windows icons is 38,4%. These unsatisfactory recognition rates could not ensure the effective machine operations in China, if the information is coded only through symbols or icons.

Figure 10.4 presents the symbol recognition results arranged according to their abstractness. 6 pictorial symbols and 6 abstract symbols are presented. The average recognition rate for pictorial icons is 61,3% and for abstract icons is 38,7%. This means that the recognition ability of Chinese machine operators for pictorial symbols and icons is much better than that for the abstract one.

Through ANOVA analysis the expected significant statistical difference between these two groups has not been found. This means that the grade of abstractness of a symbol would not be the decisive factor to influence the recognition ability of the Chinese machine operators. In fact, from the test it can be seen that some abstract symbols have a very high recognition rate but some pictorial symbols have very low recognition rate.

It is assumed that there should be other factors which have significant influence on recognition ability of the Chinese machine operators. Further

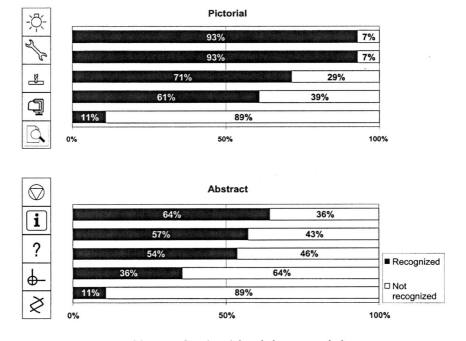

FIG. 10.4.   Recognition rate for pictorial and abstract symbols.

analysis has revealed one important characteristic which has significant influence on the recognition ability of the Chinese machine operators. This characteristic can be defined as the context degree of a symbol on its application areas and helping persons to be correctly interpreted. The *general* symbols have *high* dependence on the contexts. Their interpretation depends largely on the application area and the interpreter. The *special* symbols have *low* dependence on the contexts. Their interpretation is more definite and stabile and less dependant on the interpreters and often used in a specific application area.

The average recognition rate of the above selected symbols is then calculated according to this characteristic. The result is shown in Fig. 10.5. The average recognition rate for special symbols is 67.9% and for general symbols is 32.1%. Further statistical analysis (ANOVA) has also shown that the context dependence of the symbol has clearly significant influence on the symbol recognition results for the Chinese machine operators. This has revealed another important application characteristic of symbols on machine user interface.

Based on the above mentioned test results, for the application of symbols and icons for information coding attention should be paid to the following points:

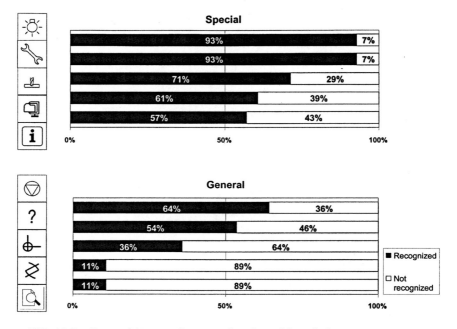

FIG. 10.5.   Recognition rate for general and special symbols.

- The application scale of icons, symbols on machine user interface in China should be restricted. For the most important and critical machine information, other information coding ways, at best the Chinese text label should be applied together to ensure the correct interpretation of the information;
- The application of abstract symbols should at best avoided in China, especially those symbols which are to be used to present important machine information. In the cases that the application of abstract symbols is necessary and unavoidable, for example according to Chinese mandatory standards, etc., the additional information coding ways, for example the Chinese text label should also be applied to ensure the correct information interpretation;
- The application of pictorial symbols and icons can not always ensure the correct information interpretation. Even the interpretation of realistic objects can be largely culture dependent. Many pictorial symbols have meanings which are too general but not specific and can be differently interpreted in different contexts. Therefore, for a specific application case, the meaning of the symbols should be further investigated.

CONCLUSIONS ON HUMAN–MACHINE SYSTEM DESIGN FOR THE CHINESE MARKET

As the intercultural human–machine system design is essentially a very complicated work for any foreign machine producers, it is necessary to provide one "general guide" which has described clearly the importance of different machine design issues in the whole localization process to let the machine producers decide on their own localization design strategy toward the target market (Röse, 2002).

The investigation revealed that the user's requirements on different machine design issues have obviously different intensity. Therefore, the localization of different human–machine system design issues on the Chinese market has different efficiency. According to the localization cost and the localization effect, all the previous mentioned design issues can be organized into three levels, which are shown in Table 10.7.

The localization issues on *surface* level concerns mainly some cultural conventions on surface, such as the language aspects, information coding standards, widely accepted formats, etc., which are normally very explicit and easy to be found in some standards and norms. The elicitation of required design features do not require large expenses from the producers, and because the design features of these issues remain usually quite stable among different application cases, the whole localization cost for these issues is normally the lowest. On the other hand, since these issues are the

TABLE 10.7
Three Levels of Human–Machine System Design Issues

Design Levels	Design Issues
Interaction	General structure issues
	Dialogue design issues
	User support issues, etc.
Function	Machine functionality
	Technological features
	Service issues, etc.
Surface	Information presentation issues
	General layout issues
	Language issues, etc.

surface of the machine, they will very easily give the target user the impression whether the machine is specially localized for them or not. And, because these issues involve mainly the information presentation on the machine, they are critical for the users to well understand the machine application. These issues have a very high localization effect when the application productivity, safety, etc. are considered. For most of the application cases, the localization of the issues on the surface level could ensure obtaining good localization effects with low cost, therefore these issues have the highest efficiency and are always localized at first.

The issues on *function* level reflect the main motivation of user's purchase to the machine. They are the prerequisites of effective machine application on the foreign market. The well addressing of user's particular requirements on these issues could largely promote the user's acceptance to the machine, namely these issues have high localization effect. The user's requirements on these issues are not difficult to be elicited (through some market research methods), however, the adaptation design issues on function level could involve the major change of machine functionality or even the significant machine design and production, which could increase the localization cost significantly. Therefore, the localization decision on this level needs careful comparison of the cost and effect and is usually implemented only when the functional features of the machine are not applicable on the target market without the adaptations. Sometimes for some issues on this level the localization efficiency could be very high and it could also be implemented in the early localization phase.

The issues on the *interaction* level are closely related to the cultural conventions "below the surface," namely thinking patterns, problem solving strategies, etc. of the users. Although they have large influence on user's effective interaction with the machine and are quite important for the machine application, however, because of the high difficulty to elicit target user's characteristics on this level, the localization cost for these issues would be much to high. These issues are at present being analyzed by more and more researchers and after there are more applicable design guidances the producers could also gain much better from the localization on this level. But at present, the practical localization on this level will meet much difficulty.

The localization efficiency which has been explained above can be illustrated by Fig. 10.6. The investigation results in China have well demonstrated this basic localization strategy. The localization of the design issues on the surface level (most significantly on interaction language) and on the function level (especially on machine quality and service) there should be more emphasize than other design issues.

It is quite clear that the intercultural human–machine system design is one good warranty for German machines to be well accepted on the Chinese market. The well address of culture-specific user requirements could ensure the benefits of the German machine producers in China. In this paper, the most important and significant user requirements on imported machines on the Chinese market are analyzed and summarized in a way to make them to be as close to the objective facts as possible. The machine localization design issues and their design features for the Chinese market have provided a solid basis for the German machine producers to conduct their machine localization design practice in China.

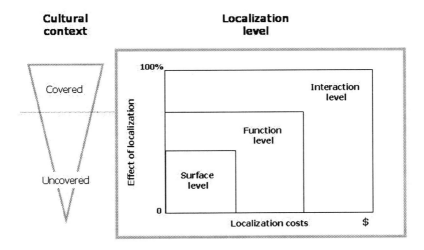

FIG. 10.6.   Localization efficiency of different localization levels.

All in all, the project INTOPS-2 has constructed one general intercultural human–machine system design structure for the Chinese market. Further research is still needed to formulate more detailed design issues and design approaches to conduct the machine localization design in practice. In the near future, a more detailed design style guide would be the best way to facilitate the developers to localize their machines more effectively for the Chinese market.

ACKNOWLEDGMENTS

Many thanks to the sponsors of this project, including MAN Roland Druckmaschinen AG, IWKA AG, Bühler AG Amriswil, and Rittal Rudolf Loh GmbH & Co. KG. A special thanks to the PhD student Long Liu, who collected the data in Mainland China and worked with high engagement on this project and this publication.

REFERENCES

Bannon, L. J., & Bodker, S. (1991). Beyond the interface: Encountering artifacts in use. In: J. M. Carroll (Ed.), *Designing interaction: Psychology at the human-computer interaction.* (pp. 227–253). Cambridge, UK: Cambridge University Press.

Choong, Y.-Y. (1996). *Design of computer interfaces for the Chinese population.* Unpublished doctoral dissertation, Purdue University, West Lafayette, IN.

Choong, Y.-Y., & Salvendy, G. (1998). Design of icons for use by Chinese in Mainland China. *Interacting with Computer, 9*(4), 417–430.

Del Galdo, E. M., & Nielsen, J. (1996). *International user interfaces.* New York: Wiley.

Dong, J. M., & Salvendy, G. (1999). Designing menus for the Chinese population: Horizontal or vertical? *Behavior & Information Technology, 18*(6), 467–471.

Fernandes, T. (1995). *Global interface design: A guideline to designing international user interfaces.* London: Academic Press.

Hofstede, G. (1997). *Cultures and organizations: Software of the mind.* New York: McGraw Hill, 1997.

Hoft, N. (1996). Developing a cultural model. In E. M. del Galdo & J. Nielsen (Eds.), *International user interfaces* (pp. 41–73). New York: Wiley.

Honold, P. (1999). Cross-cultural usability engineering: Development and state of the art. In H.-J. Bullinger & J. Ziegler (Eds.), *Proceedings of the 8th International Conference on Human-Computer Interaction. Human–Computer Interaction: Ergonomics and user interfaces* (Vol. 1, pp. 1232–1236). Mahwah, NJ: Lawrence Erlbaum Associates.

Marcus, A. (2001). Cross-cultural user-interface design. In M. J. Smith, & G. Salvendy (Eds.), *Systems, social and internationalization design aspects of human–computer interaction.* Volume 2 of the Proceedings of HCI International.

Plocher, T. A., Garg, C., & Chestnut, J. (1999). Connecting culture, user characteristics and user interface design. In H.-J. Bullinger & J. Ziegler (Eds.), *Proceedings of the 8th International Conference on Human–Computer Interaction. Human–computer interaction: Ergonomics and user interfaces* (Vol. 1, pp. 803–807). Mahwah, NJ: Lawrence Erlbaum Associates.

Röse, K. (2002). *Methodik zur Gestaltung interkultureller Mensch-Maschine-Systeme in der Produktionstechnik* (Method for the design of intercultural human-machine systems in the area of production automation). Fortschritt-Berichte pak Band 5, Kaiserlautern, Verlag Universität Kaiserslautern.

Russo, P., & Boor, S. (1993). How fluent is your interface? Design for international users. In *Proceedings of INTERCHI '93* (pp. 342–347). Amsterdam:

Shih, H. M., & Goonetilleke, R. S. (1998). Effectiveness of menu orientation in Chinese. *Human Factors, 40*(4), 569–576.

Stewart, E., & Bennett, M. J. (1991). *American cultural patterns: A cross-cultural perspective*. Yarmouth, ME: International Press.

VDMA (Verband Deutscher Maschinen- und Anlagenbau). (2002). *Maschinenbau in Zahl und Bild*. Frankfurt: Maschinenbau Verlag GmbH.

Zühlke, D., Romberg, M., & Röse, K. (1998). Global demands of non-European markets for the design of user-interfaces. In *Proceedings of the 7th IFAC/IFIP/IFORS/IEA Symposium. Analysis, Design and Evaluation of Man-Machine Systems* (pp. 000–000). Kyoto, Japan:

Chapter 11

Travel Planning on the Web: A Cross-Cultural Case Study

Helmut Degen
User Experience Consultant, Düsseldorf, Germany

Kem-Laurin Lubin
Siemens Corporate Research, Princeton, NJ

Sonja Pedell
University of Melbourne, Victoria, Australia

Ji Zheng
Technische Universität Ilmenau, Ilmenau, Germany

INTRODUCTION

The Role of Economy of Cross-Cultural User Interface Design

Siemens is a global player employing more than 417,000 people in 192 countries. To be economically viable, it is a condition to offer products and services that meet the needs of people and potential clients within the global target markets. The quality of a product is perceived, to a considerable extent, by its user interface, indicating that the user interface of a product plays a key role in a successful commercialization.

Recent studies show that people in different cultures have different requirements and needs regarding the design of user interfaces (Evers, Kukulska-Hulme, & Jones, 1999; Honold, 2000; Prabhu & Harel, 1999).

From an economic perspective, there are two conceptual extremes in international marketing strategies. One is internationalization,[1] by which every country and culture receives the identical user interface. The other extreme is localization, by which every culture receives a culture-specific user interface.

These theoretical extremes have a direct impact on the cost and volume of sales. In the case of internationalization, the development costs are low. The volume of sales might be low because the user interface very often does not have a broad user acceptance. In the case of localization, the development costs are high and the corresponding volume of sales is likewise higher because the user interface has a broader user acceptance within its specific culture. One point of view is to attain 100% internationalization to keep product costs low; the other point of view is to maximize the revenue by 100% localization. To be economically successful, an appropriate balance of internationalization and localization is necessary.

However, the challenge is to determine how much adaptation is necessary to meet the needs of the target market while maintaining an optimal balance between costs and revenue. To answer this question, we conducted the following case study.

Objective and Scope of the Case Study

In our case study, there are two questions in the context of cross-cultural "User Interface Design": First, how great are the cultural differences and what kind of localization is needed for user interfaces in different cultures? Second, where, within the entire design process, should cross-cultural aspects be taken into consideration? To answer these questions, the underlying objective of our case study is to identify differences and similarities of user interfaces and how the environment of the cultural regions represented in this study impacts these differences and similarities. We propose a set of design factors to isolate these differences and similarities systematically. The target cultures selected are those of the United States, Germany, and China, more specifically New Jersey, Munich, and Beijing, respectively. These three locations were chosen because Siemens Corporate Technology "user interface design" departments are located in these three countries, making concurrent studies and communication easier to manage.

In selecting what market domain would best suit our study, it became apparent that we had to select an area that fulfilled the following requirements: It should (a) play a major role in each culture, (b) be deeply anchored in the local culture, and (c) be easy to find participants for

---

[1]Internationalization is defined here as the production of only one product version for the whole global market. Therefore, every country and culture receives the identical user interface.

conducting the case study. Therefore, we chose the market domain of travel—an area that is entrenched in most cultures. More specifically, we chose the task of "travel planning" in a web-based travel portal for private consumers. The idea is that in contrast to job-relevant topics we have a broader communication basis that prevents misunderstandings with the participants as well as among the three different research parties in United States, Germany, and China.

Another important criterion was to select the appropriate target user group. The target user group selected for this study is "double income, no kids" (DINKs). This target user group is easily identifiable across cultures. This group is characterized as having a lot of freedom with respect to time and money. This means that the focus is on the behavior of the users and user interface itself and not on the process or methods that lead to the interface. Therefore, these commonalities enable a more standardized and equal comparison.

In this paper, we often refer to Americans, Germans, and Chinese. Our references to these groups do not cover the entire population of each country, but only the participants in this case study, as representatives of each country and culture.

Cultural-Dependent Differences and the Design Process

The first question investigated was the differences among the three user interfaces for the respective countries. In the event that there were differences, the task was to classify and compare the results of the different cultures. To guide the comparison process, we use design factors (content, function, layout, linkage, wording, and media) that allows us to assess the specific elements of interface designs (Degen & Pedell, 2003). However, in this cross-cultural study, we focus primarily on four of six design factors: content, function, layout, and linkage; we do not explicitly deal with wording and media. The reason for omitting the design factor, "wording," is that we had to complete a comparable linguistic analysis of three languages: German, United States English, and Chinese, in terms of colloquialisms and expressions used in the graphical interface. We did not have linguistic experts available at this time for this specific analysis. The design factor "media" is discussed implicitly as part of the design factor "layout." This was in part due to the fact that apart from different visual media such as text and graphical elements visualized in the single web pages, there were no other media such as sound and animations.

1. Content: Which kind of content do the users expect? How should the contents be clustered?
2. Function: Which kinds of functions do the users expect? To which content clusters are the functions assigned?

3. Media: In which media form (e.g., text, graphics, sound) should content and/or function be presented?
4. Wording: In which (textual, graphical, acoustical) expression form should the content and/or function be denoted?
5. Layout: How should the single user interface elements be organized on a single web page? How should they be designed visually?
6. Linkage: Which is the expected sequence of single web pages?

Where in the design process do the differences appear? The general question is where in the design process cross cultural design should be located. Is it only at the end of the design process (translating wording) or is it rather at the beginning? It is possible to locate the differences of cross cultural design in the design process with the introduced design factors. The design factors do not identify the type of difference but also where the information appears within the design process. We can namely allocate each design factor to a respective phase in the design process. For this purpose, we use the design process model based on the ISO 13407, a standardized Human-Centered Design Process for Interactive.

System.   The following list shows which phase of this design process refers to which design factor.

- Use Context Analysis: Content, function
- Requirements Analysis: Wording, Layout, Media, Linkage
- User Interface Design: Layout
- Evaluation

Usually within the use context analysis, user interface designer obtains information about the required contents and functions of applications. The requirements analysis uncovers expected wording, layout, media, and linkage information. During the user interface design, the user interfaces are realized in terms of interaction and visual design. During an evaluation phase, usability issues are identified. To answer the question of where cross cultural differences appear within the design process we also can look at the relationship between design factors and phases of the design process the other way around. If we identify differences regarding the design factors, e.g., content, this could already be identified during the user context analysis. Another example: If we identified differences regarding the layout it can already be gathered during the requirements analysis.

APPLIED METHODS

Process Overview

The entire design process, until the final design of user interfaces was closely observed to investigate the drawn-up questions. We focused on a use context analysis, requirements gathering, and user interface design according to ISO 13407.

The use context analysis is seen as indispensable for obtaining knowledge of the background conditions of each culture. These conditions are assumed to have an influence on the interface design. During the requirements gathering phase, the interface qualities are created. At the same time, the results of this step bridge the gap between the use context analysis and the next phase: the user interface design.

The study was very carefully organized such that each of the three countries used identical methods and followed the same procedure (including participants, target user group, length, used material, etc.). Therefore, we applied methods from a design process developed for web-based e-business applications, called JIET (Just in E-Time; Degen & Pedell, 2003) in all three countries.

Use Context Analysis

Objective

The use context analysis was the starting point. The goal of this step was to uncover the typical travel behavior of the target user group and the role of the Internet within the travel planning process.

Questionnaire

For gathering information, a questionnaire concerning travel behavior and some demographic data of the participants was used. This questionnaire allows for the gathering of day-to-day cultural idiosyncracies that would otherwise be undiscovered. Factors that might influence travel behavior (e.g., income and leisure time) were also investigated. The exact content and results of the questionnaire are shown in the result section of this chapter.

Requirements Gathering

Objective

The objective of the requirements gathering is to identify use scenarios and user interface requirements.

Focus Group

For collecting the requirements, focus groups were conducted. Each of the focus groups in the different countries had four participants who were representatives of the target user group, DINKs. Within each focus group, the following method was used.

Paper Prototyping Method NOGAP

The objective of the requirements gathering is the detection of use scenarios and user specific requirements for the user interfaces. The requirements-gathering step bridges the gap between the analysis and the user interface design (Wood, 1998). To this extent, it is essential, for efficiency reasons, to arrive as close as possible to the desired design of the user interfaces. Therefore, at this point, we used a specially developed paper prototype procedure named NOGAP (*An*other Method for *Ga*thering Design Requirements in E-business User Interface Design *P*rojects; Degen & Pedell, 2003). At the core of the procedure are the six design factors mentioned earlier in this chapter. This categorization is used on a form for requirements gathering (Fig. 11.1) and a form on which users outline their ideas (Fig. 11.2).

During this entire data collection step, the user only works with the sketch form. The user develops and sketches ideas and wishes and is guided by the user interface designer. The result is a direct image of the user's mental model (Pedell, 1998). Even users with little design experience begin to sketch their ideas after a short warm-up phase (approximately 5 minutes). The user interface designer records the ideas on the requirements form, which he or she also uses as guideline for later questions and clarifications. It is recommended to carry out this type of interview procedure with two user interface designers, where one designer asks the questions; the other records the requirements. In this situation, a transparency of recording the results is especially important. The simplest and most efficient way is to record directly with a laptop computer. During the entire time, all participants can view the records shown via projection. This procedure increases the confidence between the persons involved, and at the same time it prevents incorrect recordings. The results of the requirements gathering are used as a basis for the design of the user interface.

Scenario				
Scenario No. and Name:			**Step No. and Name:**	
Webpage				
No. (previous webpage)	**Webpage No.:**		**Heading of the webpage:**	**No. (following webpage):**
	What is the intention of the webpage:		**Decision relevant information:**	

User acitivities:			**Requirements:**
			    Content:
			
			    Media (e.g. text, icon, sound):
			
			    Wording:
			
			    Layout:
			
			    Links (with wording and no. of following pages):
			
			    Functions:	

Examples				

FIG. 11.1.   Requirements form.

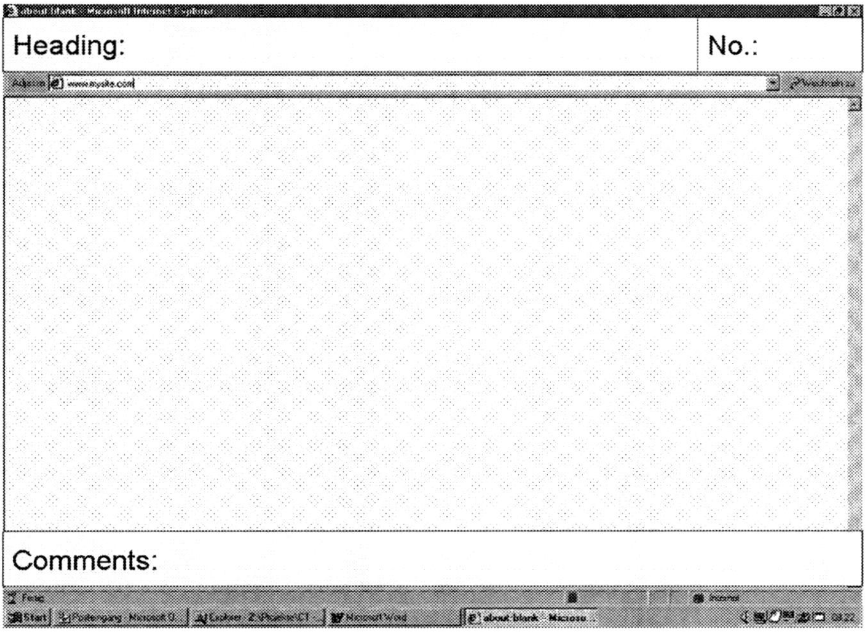

FIG. 11.2.   Sketch form.

For our study, we documented the click-through parts for the scenarios. Later on, we compared the results of two groups of one country to identify the major design ideas and elements for one common country design.

User Interface Design

Based on these ideas and elements, the user interface designers created a web site concept including the requested content, layout, and click-through. This concept was delivered to the visual designer in the form of a grid design (sketched with Microsoft Visio). In a design workshop, we explained the concept and the background information to the visual designer. He or she then had the task and freedom to shape the interface concept into an attractive design, with two restrictions: considering the given results and the respective target culture.

RESULTS

Use Context Analysis

For interpreting differences in cross-cultural design, it is very important to know about the users' circumstances. Next, some general aspects of the countries and participants' lives are discussed. These aspects were chosen for their probable influence on travel behavior and differences in the three countries.

Currency and Buying Power

In particular, financial background influences both everyday life and travel behavior.

The buying power of the currencies differs dramatically among the three countries (see Table 11.1). The dollar has eight times more value than the Chinese RMB; this means that the Americans get the value of eight dollars for every one dollar when traveling within China. On the other hand, the Chinese have to pay eight times more in the United States than in their own country. The value of the Euro compared to the RMB is similar. The value difference between the Euro and the U.S. Dollar is not as dramatic, but still 1.4 in favor of the Americans.[2]

Target User Group: DINKs

As mentioned before, we chose the target user group DINKs (Double Income, No Kids). All three countries share this target user group. Addi-

---

[2]Based on exchange rates at the time of study.

TABLE 11.1

Currency and Buying Power

	Germany	United States	China
Currency	Euro and Cent	Dollar and Cent	RMB and Fen
	1 Euro = 100 Cents	1 US Dollar = 100 Cents	1 CNY = 100 Fen
Conversion Rate	1 Euro = 0.9 US Dollar	1 US Dollar = 1.1 Euro	1 CNY = 0.13 Euro*
	1 US Dollar = 4 CNY*	1 US Dollar = 8 CNY*	1 CNY = 0.12 US Dollar*
Buying Power	Euro in United States: 70%	US Dollar in Euroland:[1] 140%	RMB in Euroland: 13%*
	Euro in China: 400%*	US Dollar in China: 800%*	RMB in US: 12%*
Sources	Statistisches Bundesamt 2002	Statistisches Bundesamt 2002	State Administration of Foreign Exchange
	*Estimated	*Estimated	*Estimated

[1]Euroland is the region in Europe with the Euro as the valid currency. The Euroland consists of the following countries: Austria, Belgium, Finland, France, Germany, Greece, Ireland, Italy, Luxembourg, The Netherlands, Portugal, and Spain.

tionally, this group was an appropriate selection because of their unlimited financial possibilities and unrestricted ways of planning their travel.

Table 11.2 shows some fundamental similarities across cultures, but related to buying value, the Americans have more freedom in spending their money than do the Germans. The Chinese DINKs earn a lot more money than the average non-DINK Chinese class, but less than the Americans and Germans. The age and the status are similar across all countries. Concerning their linguistic ability, the American participants usually do not speak a foreign language. While not of primary interest for the purpose of this study, the languages factor may influence their travel behavior as well.

Participants

The participants are representative of the target user group, DINKs, which allows for a more reliable and comparable result of this cultural study. Table 11.3 shows the most important demographic data about the participants. Again, we are able to ascertain some fundamental similarities in the profiled target group which serves for comparability.

TABLE 11.2
Demographics of Target User

	Germany	*United States*	*China*
Age	25–35	28–36	25–35
Status	Professionals	Professionals	Professionals
Income (2 persons)	80,000 Euro	150,000 USD	120,000 RMB
Education	High educated (often university degree)	High educated (often university degree)	High educated (often university degree)
	Multilingual (English and other languages)		Multilingual (English and other languages)
Source	Participants	Participants	Participants

Significant differences between women and men within the target user group DINKs could not be found. Of interest is that the Chinese participants spent a higher percentile income than either the Germans or the U.S. participants.

Frequency and Length of Private Travels

One study shows that one of the basic factors in choosing a travel destination is average annual leave. Available vacation time varies significantly among the three countries (see Table 11.4). Germany has the longest an-

TABLE 11.3
Participants

	Germany	*United States*	*China*
Number of persons	4 persons	4 persons	4 persons
Target User Group	DINKs	DINKs	DINKs
Gender	3 male, 1 female	1 male, 3 female	2 male, 2 female
Range of Age	30–33 years old	28–34 years old	26–29 years old
Expenses for private traveling per year (only DINKs) for 2 persons	6,000 Euro (8% of yearly income)	8,000 USD (5% of yearly income)	18,000 RMB (15% of yearly income)
Source	Participants	Participants	Participants

TABLE 11.4

Frequency and Length of Private Travels

	Germany	United States	China
Average annual leave (1 person)	6 weeks	2 weeks	3 weeks at least
Public holidays	Depending on the region Average: 11 days	11 days	10 days
Frequency of private traveling	2–7 times a year (mostly 3 times a year)*	2 per year*	1–3 times a year (mostly 2 times a year)*
Length of private traveling	3 days–4 weeks (mostly 2 weeks)*	1–2 weeks*	1 week average
Source	*Participants	*Participants	*Participants

nual leave. Only the Germans have the possibility of traveling for 3 to 4 weeks in a row. This might be a better incentive to travel far away. The American and Chinese travel time, by comparison, is much shorter, which may account for a "close-to-home" attitude in their travel.

Travel Behavior

Travel behavior differs considerably among the three countries (see Table 11.5). Whereas the Germans prefer to travel individually, the Americans and Chinese prefer to travel in groups. The Germans seem to have a far greater interest in the weather than do the participants from the other two countries. The participants from China do not have a concrete preferred travel destination. Also of interest was the primary criteria that made a destination desirable. While German and American participants cited "weather," the Chinese chose nature and landscape. Table 11.5 shows the responses by frequency.

Preferred Communication Channels

As further input for later design, we paid special attention to the mode of communication for each travel planning step. Table 11.6 shows the usual task flow for travel planning for each country. The first step is the selection of a travel destination, as well as selecting arrangements (accommodation and

TABLE 11.5
Travel Behavior

	Germany	United States	China
Manner of travel	Individual or 2 persons	With others	As a group
Destination Criteria	Weather (4)	Weather (4)	Nature landscape (3)
	Exotic (2)	Culture (4)	Culture (2)
	Sightseeing (2)	Food and Wine (3)	Nature (1)
	Culture (2)	Beach (2)	Sports: Climbing (1)
	Nature (2)	Recreation (1)	Service (1)
	Beach (1)		
	Food and Wine (1)		
	Languages: English, French (1)		
	Sports, activity (1)		
	Recreation (1)		
Preferred Destinations	Europe, America, Southeast Asia	South America, Caribbean, and Europe	Main objective: recreation as a change from the city. To see foreign countries
Booking by Internet	Yes (4)	Yes (4)	Yes (4)
Source	Participants	Participants	Participants

means of transport). Following are the steps of reservations, booking and confirmation. Sometimes travels arrangements are changed or rebooked, which requires user access to flight information. In case of waiting times, participants want to have information about the current status of delivery (tracking/tracing). For each step, the participants communicated to us their preferred communication channels. The only exact commonality is the participants' Internet and telephone preference for the final stage (tracking/tracing).

However, in the United States and in China, the Internet is one of the preferred communication channels in every step. German participants do not use the Internet, the Americans prefer the telephone, and the Chinese participants prefer the Internet. In Germany, the main reason for not using the

TABLE 11.6
Preferred Communication Channels for Travel Planning Cycle

Single steps	Germany	United States	China
Select Travel	1: In-Person	1: In-Person	1: Internet
	2: Internet	2: Internet	2: In-Person
Select Arrangement	1: In-Person	1: In-Person	1: Telephone
	2: Internet	2: Internet	2: Internet
Reservation	1: In-Person	1: Internet	1: Internet
	2: Telephone	2: Telephone	2: In-Person
Booking	1: In-Person	1: Telephone	2: In-Person
	2: Telephone	2: Internet	1: Internet
Confirmation	1: In-Person	1: Internet	1: Internet,
	2: Telephone	2: Telephone	telephone
Change/Rebook	1: In-Person	1: Telephone	1: Telephone
	2: Telephone	2: Internet	2: Internet
Tracking/Tracing	1: Internet	1: Internet	1: Internet,
	2: Telephone	2: Telephone	telephone
Source	Participants	Participants	Participants

Note. Of specific interest was the German affinity for person-to-person communication in all steps but the last. 1 and 2 refer to the primary and secondary preferred communication channels.

Internet is security. There is a hesitation and aversion to reveal their credit card information. In China, this problem doesn't exist to that extent, because credit cards are not very popular. In Shanghai, one of the biggest cities in China, the percentage of credit card holders was 3.8% at the end of 2001 (Xinhua News Agency, 2001). A very popular form of payment is "Pay on Delivery." With "pay and deliver," the Chinese user orders a flight ticket via the Internet or the telephone and gives the travel agency the contact information. The ticket will be delivered and paid by the user during the handing over; this is like the American version of collect on delivery (COD). Besides this form of payment, there are also other traditional payment methods like cash, bank remitting, and invoice which replace the more Western forms of online payment.

Summary

To summarize the use context information we can say that the similarities within one country, among the participants, are extremely high, but

TABLE 11.7
Requirements Based on Design Factors

Design Factor	Germany	United States	China
Content	**Home Page** • Information about countries and destination (travel guide) • Filter Information: Activities, weather, travel suggestions etc. • Special Offers • Transportation and Hotel Information • Price, dates, and quality information • Availability of offerings **Required Travel Information** • Departure and arrival times • Hotel address	**Home Page** • Specials • Quick filters • Link to Accommodations (hotel and car) • Useful travel resources • Travel News • Customer Service information • Expert Advice **Required Travel Information** • Departure and arrival times • Hotel address	**Home Page** • Categorized travel information such as domestic/abroad, cultural, geographic etc. • Travel news • BBS/Clubs for information exchange • No flight or hotel information, but these should be specified in the travel package offers from travel agency • Select target city and travel offers are the aims of the users while using this web **Required Travel Information** • Appointment information of the group
Function	Booking system full text search engine	Booking system quick search of flight information	Booking system full text search engine
Layout	• Left-hand side: Filter criteria • Middle: Map • Right-hand side: Offerings	• Left-hand side: Traveler tools • Middle: Search field • Right-hand side: Specials Offers	• Left-hand side: Search/Information Categories • Middle: News, Highlights, Offers • Right-hand side: special topics, add-value service
Linkage	1. Using filter criteria and region (or vice versa): Alternative: Search Engine 2. Travel Information 3. Travel Offerings 4. Travel Booking	1. Using the search engine; alternative: selecting special offers 2. Travel Offerings 3. Travel Booking	1. Selection domestic or abroad destinations; alternative: Search Engine 2. Selecting regions or special offerings 3. Travel Information 4. Travel Offerings 5. Travel Booking

some differences among the three countries are evident. Although the Americans have most financial freedom for their travel, the Germans have the largest blocks of free time to spend abroad. The Chinese participants spend 15% of their yearly income on traveling, which is the highest percentage spent of those studied. For the Chinese, traveling seems to equate high quality of life. The Germans, by choice, have the lowest Internet affinity; they prefer the direct contact with salespeople. When planning a trip, the Americans prefer contacting a salesperson, but for reserving, booking, tracking/tracing, they prefer the telephone or the Internet as communication channels. The Chinese prefer the Internet for almost every step. Only for rebooking and for travel arrangements do they prefer the telephone.

Requirements Gathering

The requirements were gathered in all three countries equally with the method described earlier in this chapter. Each focus group in all three countries was divided in two groups of two persons. These pairs came up with very similar ideas that were merged by the user interface designers in one layout in the next step. Additionally, two extra focus groups took place in the United States (within a tutorial of the CHI 2002 conference), and in China (within a demonstration workshop for this method). The same persons who moderated the German focus group for this study also moderated these two extra groups. The collected requirements cover the results presented here and support them. That means, as well, that the differences found among the countries derive from cultural differences rather than from differences in using the method.

Germany Results

Very often, the Germans do not have a fixed travel destination in mind. They want to find a travel destination based on personal criteria (e.g., weather, culture, exotic). For the German participants, searching for the travel destination is the start and preparation for the travel itself. They enjoy "browsing" around in the world. The web site is a means to collect ideas, find a travel destination, and find precise information (e.g., about availability, prices, and quality).

The Germans wish to get a quick overview of the destination possibilities in general (the whole world). Quick access to the regions of the world (globe) in the center of the screen proved to be the ideal solution. According to their travel expectations, they want to have access to filter criteria (e.g., weather, activities, social circumstances) on the left-hand side of the screen. They expect concrete offerings for the selected region and filter criteria on the right-hand side, as shown on the scribble (Fig. 11.3).

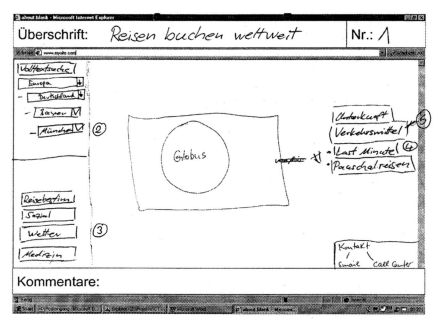

FIG. 11.3.   Home page (Germany).

Furthermore, they like to have direct access to proposed travel (e.g., last minute offers, all-inclusive packages) on the right-hand side as well. In addition, they need clear access to contact information (see Fig. 11.3: lower right corner).

After selecting a region (in this case, Munich), the participants expect to see a description very saliently placed on the page (see Fig. 11.4). On the left-hand side of the page remains the filter area and on the right-hand side the access for concrete offerings (hotel, transportation, last minute offers).

To classify the differences among the three user interfaces in more detail, their characteristics were allocated to the design factors described earlier.

We summarize the results of the German requirements using the following four design factors of content, layout, linkage, and function, respectively.

Content.   The first screen shows the globe and the second screen shows the destination, in this case, Munich. Filters represent criteria by which the German participants choose their travel destination, such as weather conditions, medical situation in the country, and cultural life. The filters stay on the home page and filter further within the chosen area in more detailed destination levels (Munich, in this case).

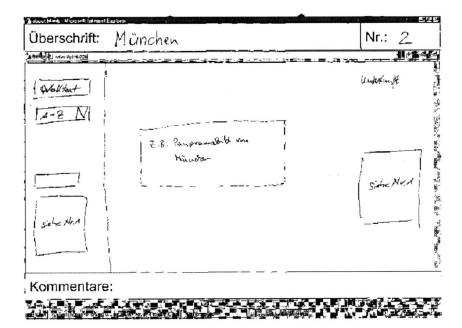

FIG. 11.4.   Destination (Germany).

Layout.   Content and visual information go very close together. It is very important that the layout gives a clear overview and supports the user's tasks. There is a direction from left to right that allows the user to filter the travel information: filter criteria (left column), region (middle column), and offerings (right column). (See Fig. 11.3)

Linkage.   As sketched in the interaction architecture in Table 11.8, the users mainly prefer three different ways of finding their travel information: (a) using the filter functions and then the regions, and (b) using the regions first and then the filter functions, and (c) using the search engine. After getting the travel information, the users select relevant travel offerings and might subsequently book the travel.

For example, in navigation path 1, the user starts on the home page, then applies travel filter (e.g., activity travels) (area 2 on Fig. 11.3) and selects the travel destination by a region (area 3 on Fig. 11.3). Afterward, the user receives travel information and can get travel offerings about the selected region. Finally, travel arrangements can be booked.

TABLE 11.8
Interaction Architecture for Germany

Web page	Navigation Path 1	Navigation Path 2	Navigation Path 3
1	Home Page	Home Page	Home Page
2	Travel Filter	Region	Search Results
3	Region	Travel Filter	Travel Filter
4	Travel Information	Travel Information	Travel Information
5	Travel Offerings	Travel Offerings	Travel Offerings
6	Travel Booking	Travel Booking	Travel Booking

Note. Reading example for navigation path 1: The user starts on the home page (1) and applies travel filter (e.g., activity travels) (2) and selects the travel destination by a region (3). Afterward, the user gets travel information (4) about the selected region and can get travel offerings (5). The user can book these (6).

Function. After finding the right region for travel, the participants want to have access to reservation tools if they choose to use them. This is a step toward getting a concrete and real idea of their complete travel plan and to see the availabilities. They still prefer to leave the booking itself to the agency.

U.S. Results

In contrast to the German participants, the American participants have a very concrete idea of their travel destination. In part, this is due to the short vacation times, compared to their German counterparts who have much longer periods for travel. It is safe to say that most of the information gathering is very much in the background and in some cases nonexistent. Our studies show that American participants want to have a very effective and efficient tool to book their travel that takes into consideration time constraints. Therefore, the scribble of the home page shows different boxes, which have all the intention to provide concrete information for somebody who knows exactly what he or she needs for a certain travel. A clearly placed "Quick Search," in the center of the home page, and the main categories, such as air, hotel, and so on, placed in the top tabs show this intention clearly. For users who haven't chosen their destination, the web site provides special offers (today deals) and "destination of the week," which both include ready travel packages for the intended travel.

We summarize the results of the U.S. requirements using the same design factors.

Content. The tab cards remain in view throughout the whole Web site navigation and allow for quick change from one category to the next. All aim directly at the easy booking. At the second screen, when navigating from the main screen (Fig. 11.5), a concrete flight can be booked as shown in Fig. 11.6.

Layout. A layout is chosen, which corresponds with this booking-oriented content. The content is offered in portletlike packages that contain different kinds of information. The quick search is the central element and takes most of the central space (Fig. 11.5).

Linkage. As shown in the interaction in Table 11.9, the users prefer two different ways to get to the travel offerings. One path uses the search engine and the other path is through the search for special offerings. Following, travel offerings are displayed and the users can book their travel arrangements.

For example, for navigation path 1: The user starts on the home page and proceeds to using the search engine. After getting the search results, travel offerings can be selected and are displayed according to search filters. Then selected offerings can be booked.

Function. The functions are all related to reservation, booking, and getting detailed information about concrete travel destination. These con-

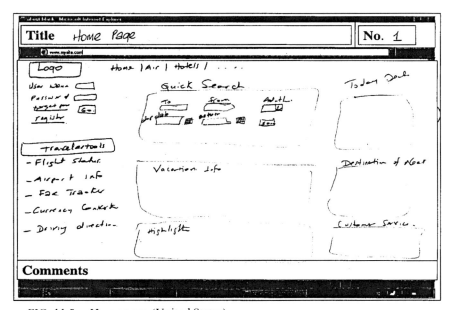

FIG. 11.5. Home page (United States).

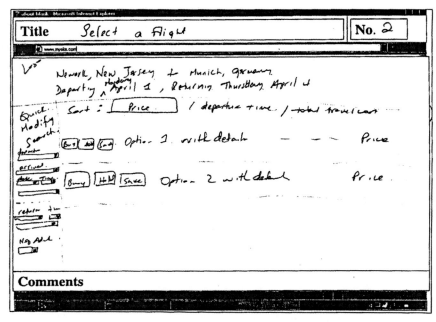

FIG. 11.6.   Destination (United States).

TABLE 11.9
Interaction Architecture for the United States

Web page	Navigation Path 1	Navigation Path 2
1	Home Page	Home Page
2	Search Results	Special Offerings
3	Travel Offerings	Travel Offerings
4	Travel Booking	Travel Booking

crete functions, in contrast to information gathering are on a very top level of the web site. This means that users encounter the concrete and primary function on arriving at the main site. Additionally, the quick search provides the possibility to choose the means of transport.

China Results

The Chinese participants seem to be easily enticed by pictures that show different travel destinations. They pursue a high quality of life and their disposable income essentially affords for inclusion of traveling.

Even though traveling overseas is very attractive to them, one main focus is on their own country. The reasons for this might be the short holiday time, the size of their country (China provides a large variety of landscapes and different cultures), and the organizational effort (e.g., acquiring a visa) that is needed for traveling abroad.

Content.   The first scribbles (Fig. 11.7) shows the following contents: a search engine, the most traveled tourism cities and cultural places, special customs, recommendations, tourism-relevant news, forum for information and experience exchange, and contact. The second scribble, destination China (Fig. 11.8), still includes the search engine. It also provides categorized information about the destination, such as geography, history, culture, sightseeing, customs, and many pictures and links to travel offers.

Layout.   In the home page China drawing (Fig. 11.7), the information is clearly categorized. Similar information is combined, such as added value service (SMS, travel clubs). In Fig. 11.7, the top navigation remains stable on the page. Photos and information about customs, as well as cultural de-

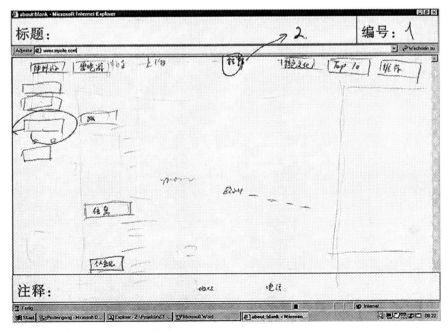

FIG. 11.7.   Home page (China).

tails, appear in the middle of the page. Categorized information and search engine appear on both sides.

Linkage.   On the first scribble (Fig. 11.7), users want to select whether they plan to travel domestically or abroad. Following, lists or categories of destinations are presented. The most popular tourism countries, regions, and cities are selected on the top navigation. Besides, users can also select the "Top10" to view the most traveled or recommended places. In addition, a powerful search engine, with advanced search categories like history, culture, geography, and so on, enables users to find information as quickly as possible. The idea of the second scribble (Fig. 11.8) is to select categories such as geography, culture, and sightseeing for detailed explanation. The click on "More Photos" leads to the photo gallery. If the users are interested in the destination, they can immediately select respective travel offers. The top navigation and search engine on the right side help users to find other destinations.

The users prefer four options (see Table 11.10) of completing their task. The first and the second options guide the users through the domestic destinations, the third and the fourth options through destinations abroad. In the fifth navigation path, the users use the search engine. In the case of navigation paths 1 and 3, the users prefer to select offerings according to regions and in navigation paths 2 and 4, according to special offerings. Afterward, travel information is displayed; the user can then see the travel offerings and book selected offerings.

As depicted in Table 11.10, the user starts on the home page and selects domestic travel destinations (area 2 on Table 11.10). Afterward, he or she has the opportunity to select regions or special offerings (area 3 on Table 11.10). The travel information for region or specials offerings can be selected and are displayed (area 4 on Table 11.10). The user can see available offerings (area 5 on Fig. 11.10) and can book the travel (6).

Functions.   The Chinese appreciate a powerful and categorized search engine. The following functions, which are not realized in this case study, are of special interest: Chinese participants would like to have a voice interface, for instance, they like a friendly voice to welcome them. Additionally, they are very interested in 3D virtual tour.

Requirements Gathering Summary

Based on the results of the requirements gathering, the main differences between the three countries relate to content and linkage. While the Germans and the Chinese require comprehensive travel information, the Americans do

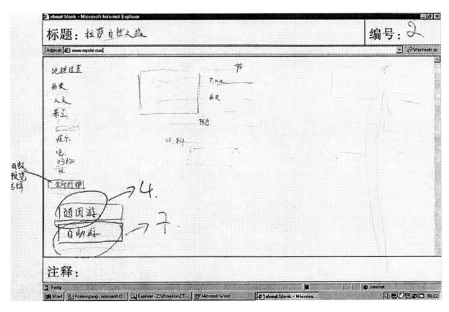

FIG. 11.8.   Destination (China).

TABLE 11.10

Interaction Architecture for China

Web page	Navigation Path 1	Navigation Path 2	Navigation Path 3	Navigation Path 4	Navigation Path 5
1	Home Page	Home Page	Home Page	Home Page	Home Page
2	Domestic	Domestic	Abroad	Abroad	Search Results
3	Regions	Special Offerings	Regions	Special Offerings	
4	Travel Information	Travel Information	Travel Information	Travel Information	Travel Information
5	Travel Offerings	Travel Offerings	Travel Offerings	Travel Offerings	Travel Offerings
6	Travel Booking	Travel Booking	Travel Booking	Travel Booking	Travel Booking

not. Also, the linkage concept is different among the three countries: Before Germans select travel offerings, they expect comprehensive travel information. The Americans prefer a straightforward strategy to travel offerings without travel information. The Chinese first select between a domestic or foreign destination; afterward, they follow a similar strategy as the Germans. All-important results of the requirements gathering are shown in Table 11.7.

User Interfaces

On the basis of the scribbles, which include the requirements of the participants of the respective country for a travel Web site, a visual design was created. A visual designer in each respective country did the final visual design. The procedure is described earlier in this chapter. The designs show the required contents and functions and basic layout.

INTERPRETATION OF CROSS-COUNTRY RESULTS

Similarities

Use Behavior: All cultures use the Web to plan and arrange travel, and for tracking/tracing their orders. Reserving, booking, confirming, and changing/rebooking are steps that participants of all three cultures take, but the Internet is not equally used for it.

Content: Travel offerings are seen as products, which are described and available on the Internet, even if the preferred destinations are different. All web sites provide special offers that range from complete travel packages to self-organized trips.

Functions: On every web site, a search engine and a log-on are available.

Layout: The chosen basic color on all designed travel web sites is blue, and tab cards are used to organize the content.

Linkage: Access to different offerings (hotel, car, flights, etc.) and destinations is available. The goal is either to get a detailed overview of a possible trip or to book a complete travel package.

Differences

Travel Behavior

Germany. The German participants prefer solo traveling to foreign countries, even to exotic places. The duration of one trip per year is longer

Germany

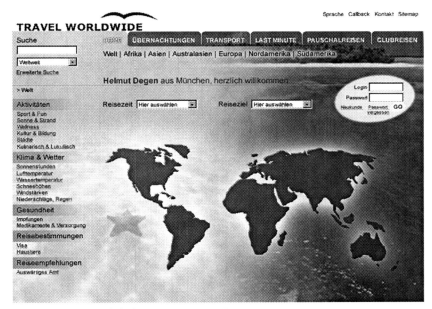

FIG. 11.9.   Visual design: Home page (Germany).

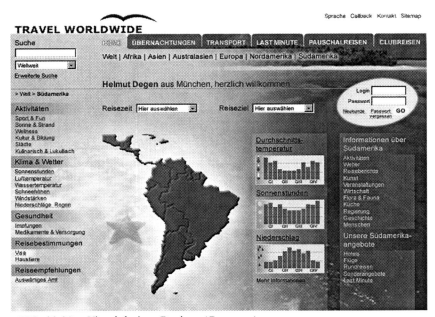

FIG. 11.10.   Visual design: Regions (Germany).

United States

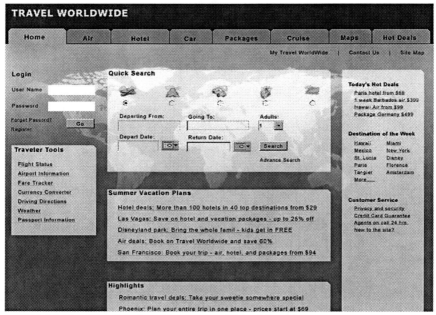

FIG. 11.11.   Visual design: Home page (United States).

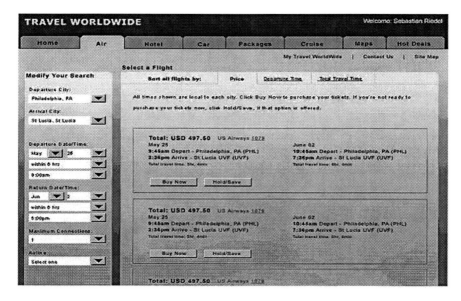

FIG. 11.12.   Visual design: Flight information (United States).

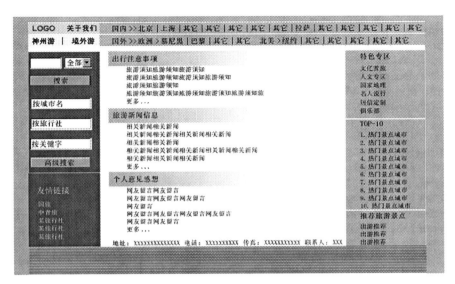

FIG. 11.13.   Visual design: Home page (China).

FIG. 11.14.   Visual design: Destination information (China).

than 2 weeks—rather 3 to 4 weeks. The Germans take their time to prepare and select their travel destination. Surfing on the Web for the "perfect spot" is fun for them. It is supposed that the fun and desire for very detailed planning beforehand goes together with the fact that the Germans travel on their own. Like the U.S. and Chinese participants, the Germans use the Internet for finding a travel destination, arranging the travel, and tracking/tracing, but not for reserving, booking, and paying.

United States.   Our study shows that more often than not, American travelers already know their travel destination when they go on the travel web site. Their choice of destination is often based on advertising and posted specials. They like traveling in pairs or in small groups with people they know. Because of the restricted travel time, they more likely travel near the U.S. continent. The Internet is used for the entire planning process except for choice of destination.

China.   The Chinese travel in groups. Individual travel is rather unusual. Like the Americans, the Chinese enjoy domestic travel. On the one hand, it is easier (organizational effort to get visa and restricted vacation time of two weeks) and on the other hand, the country itself has a large variety of different landscapes and cultures. Another reason is the lower income and buying power and the linguistic inconvenience. The Internet is used for the entire planning process from planning to booking. Like the Germans, they don't have a concrete focus on their travel destination, but they consider recommendations, web ads and other travelers' opinions to be useful whereas Germans use personal filter criteria.

Content

Germany.   The German web site provides lots of different travel and country information, as well as flight and accommodation information, and supports the user to get informed on a broad scale. Different categories like activities, climate, weather, and health lead to very detailed information that fit respective travel interests.

United States.   There are no content pages to surf, but access to concrete flights and hotels, services on site like rental cars, and special travel packages are available. Other travel options like cruise and vacation packages are quite salient, but the amount of content is not as high as seen in the German results.

China.   A lot of information relevant to the target destination can be found. Additionally, information not found on the German or U.S. travel web sites can be found here, for example, travel news, links to other travel agencies, and the Top 10 most traveled destinations. BBS[3] and information exchange of web users are also used to find the travel destination. That means that they don't have a concrete destination in mind, but make their travel choice rather by recommendations and web ads than cruising around and using personal filter criteria. Flight and hotel information are not the focus of users because, as already mentioned, these are arranged by the travel agency.

Layout

Germany. The overall layout triggers a desire to travel the whole world. Especially the home page provides a lot of function-free space to support unrestrained room for the users' imagination where to spend the next holiday. The single function filters, content information, and offers are clearly distinguished in separate blocks. They always stay in the same place while users click through the site. To support different ways of choosing a travel destination, different accesses (filters of different granularity, the world map, pull-down menus) are available.

United States. The U.S. home page is compact and gives clear directions of where to enter relevant information. Graphics and color are used to facilitate quick entry and to show relationships and grouping of information in a structured way. The contact information is clearly placed.

China. The Chinese web site provides very clear categories of target city information in rows. Regional-specific information, such as cultural events, are highlighted. Photographs of the target information are also a central feature of the Chinese web site and take up a lot of room.

Linkage

Germany. Our study shows that Germans travel enough that they are constantly looking for new and exciting travel destinations. Therefore, they browse through the web site to get travel inspiration. The click-through is as follows: Regional selection or topic-oriented selection before looking for a suitable flight and sometimes finish by looking for the right accommodations. Often, Germans merely choose the destination at the web site and book it in a travel agency.

---

[3]BBS is the abbreviation for Bulletin Board System. Web users can put texts, pictures, and animations on the system. These are normally shared for all registered users. In this case study, it is a place of information exchange, for example, users can tell about their travel experiences or show nice photos.

United States.   They have a clear idea where they want to go. There-fore, they type in the name of the destination in a Quick Search, which leads directly to a booking screen.

The procedure is similar to that in a classical Internet shop: First, they enter depart/destination search criteria; then, they receive output (e.g., flight options) and the option to hold or buy and the possibility to pay for the selected option.

China.   The intention of Chinese users, in visiting the web site, is to both browse and book. First they peruse regional selection or topic-oriented selection; then they focus on abundant information about the target destination. They book for group travel to the destination and trust that the organization of accommodation and means of transport is dealt with by the provider.

SUMMARY AND CONCLUSIONS

Designing user interfaces for the global market raises the question as to whether a high level of localization based on cultural needs is required. We have chosen the topic "Travel Portal" and the target user group "DINKs" for our case study. In this case study, we analyzed requirements and de-signed user interfaces for three cultures (Germany, the United States, and China) and showed where the differences appear. We used a set of six de-sign factors to discover and to compare cultural differences.

Based on the results of the cross-cultural design study, summarized in Table 11.7, we can identify major differences for the design factors, con-tent, layout, and linkage and minor differences for the design factor function. Based on the close relationship between the design factors and the phases of the design process (see chapter 1) we propose that cross-cultural differences are located in the phases "use context analy-sis," "requirements analysis," and "user interface design." Therefore, the cross-cultural design should be considered at the beginning of the design process and not at the end, such as simply translating the content into another language. According to the extent of differences, the over-all effort can be estimated and the level of localization can be determined in early stages.

According to the results, it is necessary to include user input of different cultures from the very beginning of the design process. Cosmetic changes such as translating the language or changing the colors, instituted late in the process of design or even after the fact, are insufficient. Managers and user interface designers can see the trend of cultural differences in an early stage of the design process (after applying the method NOGAP, require-ments phase). The effort of a final user interface design to evaluate differences is not necessary, since differences are identified early enough to warrant varying user interfaces. Here the values of localization are illustrated with success. Rather than assuming that one user interface, with final alterations, responds to the specific needs of all cultures, this case study allows for the inclusion of significant cultural factors that show how users perceive their tasks in the context of travel planning on the Web. While most companies do not consider the rudimentary cultural differences that are needed to ensure market success, an understanding of the locale idiosyncrasies undoubtedly proves useful in this study.☺

ACKNOWLEDGMENTS

We would like to thank Heinz Bergmeier, Nuo Tang, and Jui-Lin Dai for the visual design of the web pages. Many thanks to Pia Honold for the nice and constructive discussion. We thank Stefan Schoen for his interest and providing the required resources. Special thanks goes to Nuray Aykin for her patience.

REFERENCES

Degen, H., & Pedell, S. (2003). JIET design process for e-business applications. In D. Diaper & N. Stanton (Eds.), *Handbook of task analysis for human-computer interaction* (pp. 193–220). Mahwah, NJ: Lawrence Erlbaum Associates.

Evers, V., Kukulska-Hulme, A., & Jones, A. (1999). Cross-cultural understanding of interface design: A cross-cultural analysis of icon recognition. In E. del Galdo & G. Prahbu (Eds.), *International Workshop on Internationalization of Products and Systems: Vol. 1. Designing for global markets* (pp. 173–182). Rochester, NY:

Honold, P. (2000). *Interkulturelles Usability Engineering. Eine Untersuchung zu kulturellen Einflüssen auf die Gestaltung und Nutzung technischer Produkte.* Düsseldorf, Germany: VDI-Verlag.

ISO/IEC 13407. *Human-Centered Design Processes for Interactive Systems,* ISO/IEC 13407: 1999.

Pedell, S. (1998). *Die Bedeutung der Gestaltgesetze Ähnlichkeit und Nähe für die Gestaltung von Benutzeroberflächen.* Unpublished master's thesis, Technical University of Berlin, Germany.

Prabhu, G., & Harel, D. (1999). GUI design preference validation for Japan and China—A case for KANSEI engineering? In H.-J. Bullinger & J. Ziegler (Eds.), *Proceedings 8th International Conference on Human-Computer Interaction: Vol. 1. Human-computer interaction: Ergonomics and user interfaces* (pp. 521–522). Mahwah, NJ: Lawrence Erlbaum Associates.

Statistisches Bundesamt. (2002). Kaufkraft des Euro im Ausland. In http://www.destatis.de/basis/d/ausl/auslkkr1.htm, Statistisches Bundesamt, 13th of May 2002.

Wood, L. E. (Ed.). (1998). *User interface design. Bridging the gap from user requirements to Design.* Boca Raton, FL: CRC Press.

Xinhua News Agency. (2001). http://www.xinhua.org/fortune/2002-03/18/content_321301.htm

Appendix

BOOKS

Andrews, D. C. (1996). *The international dimensions of technical communication*. Arlington, VA: Society for Technical Communication.

Apple Computer. (1992). *Guide to Macintosh software localization*. Reading, MA: Addison-Wesley.

Axtel, R. E. (Ed.). (1993). *Do's and taboos around the world* (3rd ed.). New York: Wiley.

Axtel, R. E. (1995). *Do's and taboos of using English around the world*. New York: Wiley.

Axtel, R. E. (1998). *Do's and taboos of humor around the world: Stories and tips from business and life*. New York: Wiley.

Axtel, R. E. (1998). *Gestures: The do's and taboos of body language around the world*. New York: Wiley.

Bean, J. (2001). *XML globalization and best practices: Using XML schemas and XML-data*. Golden, CO: ActiveEducation.

Berry, J. D. (Ed.). (2002). *Language culture type: International type design in the age of unicode*. New York: Graphis.

Bishop, M. (1997). *How to build a successful international website: Designing web pages for multilingual markets at the national and international level*. Albany, NY: Coriolis.

Bishop, M. (1998). *How to build a successful website: Use the web to market your business worldwide*. Albany, NY: Coriolis.

Bosley, D. S. (2000). *Global contexts: Case studies in international technical communication*. Needham Heights, MA: Allyn & Bacon.

Brake, T., Walker, D. M., & Walker, T. (1994). *Doing business internationally: The guide to cross-cultural success*. Whitby, Ontario, Canada: McGraw-Hill.

Campbell, G. L. (1995). *Concise compendium of the world's languages*. London: Routledge.

Carter, D. R. (1992). *Writing localized software for the Macintosh*. Reading, MA: Addison-Wesley.

Czarnecki, D., & Deitsch, A. (2001). *Java internationalization*. Sebastopol, CA: O'Reilly & Associates.

Daniels, P. T., & Bright, W. (1996). *The world's writing systems*. Oxford, UK: Oxford University Press.

Darnell, R. (1997). *HTML unleashed*. Indianapolis, IN: Sams.

del Galdo, E. M., & Nielsen, J. (Eds.). (1996). *International user interfaces*. New York: Wiley.

De Mente, B. L. (1993). *Behind the Japanese bow: An In-depth guide to understanding and predicting Japanese behavior*. Lincolnwood, IL: Passport Books.

DePalma, D. A. (2002). *A Strategic guide to global marketing: Business without borders*. New York: Wiley.

Digital Equipment Corporation. (1991). *Developing international software*. Bedford, MA: Digital Press.

Dresser, N. (1996). *Multicultural manners: New rules of etiquette for a changing society*. New York: Wiley.

Dreyfuss, H. (1984). *An authoritative guide to international graphic symbols*. New York: Wiley.

Dr International. (2003). *Developing international software* (2nd ed.). Redmond, WA: Microsoft Press.

Elashmawi, F., & Harris, P. R. (1998). *Multicultural Management 2000: Essential cultural insights for global business success*. Burlington, MA: Gulf.

Esselink, B. (2000). *A practical guide to localization* (2nd ed.). Amsterdam: John Benjamins.

Fernandes, T. (1995). *Global interface design: A guide to designing international user interfaces*. San Francisco: Morgan Kaufmann.

Gillham, R. (2002). *Unicode demystified: A Practical programmer's guide to the encoding standard*. Reading, MA: Addison-Wesley.

Graham, T. (2000). *Unicode: A primer*. Foster City, CA: M&T Books.

Gudykunst, W. B., & Asante, M. K. (Eds.). (2000). *Handbook of international and intercultural communication*. London: Sage.

Hall, E. T. (1976). *Beyond culture*. Garden City, NY: Anchor Press.

Hall, P. A. V., & Hudson R. (1997). *Software without frontiers: A multi-platform, multi-cultural, multi-nation approach*. New York,: Wiley.

Hickey, T. (Ed.). (1999). *The guide to product translation and localization*. Washington, DC: Computer Society.

Hofstede, G. (1997). *Software of the mind: Intercultural cooperation and its importance for survival* (Rev. ed.). Whitby, Ontario, Canada: McGraw-Hill.

Hofstede, G. (2001). *Culture's consequences: Comparing values behaviors, institutions and organizations across nations* (2nd ed.). London: Sage.

Hoft, N. L. (1995). *International technical communication: How to export information about high technology*. New York,: Wiley.

Hutchins, W. J., & Somers, H. L. (1992). *An introduction to machine translation*. London: Academic Press .

Jones, S., Kennelly, C., Mueller, C., Sweezy, M., Thomas, B., & Velez, L. (1992). *Developing international user information*. Bedford, MA: Digital Press.

Kano, N. (1995). *Developing international software for Windows 95 and Windows NT*. Redmond, WA: Microsoft Press.

Kaplan, M. S. (2000). *Internationalization with Visual Basic: the authoritative solution*. Indianapolis, IN: Sams Publishing

Langer, A., & Kreft, K. (2000). *Standard C++ IOStreams and locales*. Reading, MA: Addison-Wesley.

Lunde, K. (1995). *Understanding Japanese information processing*. Sebastopol, CA: O'Reilly & Associates.

Lunde, K. (1999). *CJKV: Information processing*. Sebastopol, CA: O'Reilly & Associates.

Luong, T. V., Lok, J. S. H., Taylor, D. J., & Driscoll, K. (1995). *Internationalization: developing software for global markets*. New York: Wiley.

Madell, T., Parsons, C., & Abegg, J. (1994). *Developing and localizing international software*. Englewood Cliffs, NJ: Prentice Hall.

Marx, E. (2001). *Breaking through the culture shock: What you need to succeed in international business*. London: Nicholas Brealey.

McFarland, T. (1996). *X windows on the world: Developing internationalized software with X, Motif and CDE*. Englewood Cliffs, NJ: Prentice Hall.

Microsoft Corporation. (1993). *The GUI guide: International terminology for the Windows™ interface*. Redmond, WA: Microsoft Press.

Miller, S. (1996). *Understanding Europeans*. Santa Fe, NM: John .

Mole, J. (2003). *Mind your manners: Managing business cultures in the new global Europe* (3rd ed.). London: Nicholas Brealey.

Morrison, T., Borden, G. A., & Conaway, W. A. (1995). *Kiss, bow or shake hands: How to do business in sixty countries*. Holbrook, MA: Adams Media.

Morrison, T., Conaway, W. A., & Douress, J. J. (1997). *Dun & Bradstreet guide to doing business around the world*. Englewood Cliffs, NJ: Prentice Hall.

Nielsen, J. (Ed.). (1990). *Designing user interfaces for international use*. Amsterdam: Elsevier.

O'Donnell, S. M. (1994). *Programming for the world: A guide to internationalization*. Englewood Cliffs, NJ: Prentice Hall.

O'Hara-Devereaux, M., & Johansen, R. (1995). *Global work: Bridging distance, culture & time* (2nd ed.). San Francisco: Jossey-Bass.

Ott, C. (1999). *Global solutions for multilingual applications: Real-world techniques for developers and designers*. New York: Wiley.

Patai, R. (2002). *The Arab mind* (Rev. ed.). New York: Hatherleigh.

Prabhu, G. V., & del Galdo, E. M. (1999). *International Workshop of Internationalization of Products and Systems. Vol. 1. Designing for global markets*. Rochester, NY: USA Backhouse Press.

Rhind, G. R. (1999). *Global sourcebook of address data management: A guide to address formats and data in 194 countries*. Aldershot, Hampshire, UK: Gower.

Rhinesmith, S. H. (1993). *A manager's guide to globalization: Six keys to success in a changing world*. Alexandria, VA: American Society for Training and Development.

Ricks, D. A. (2000). *Blunders in international business*. Cambridge, MA: Blackwell.

Rockwell, B. (1998). *Using the web to compete in a global marketplace*. New York: Wiley.

Savourel, Y. (2001). *XML internationalization and localization*. Indianapolis, IN: Sams.

Searfoss, G. (1994). *JIS-Kanji character recognition*. New York: Van Nostrand Reinhold.

Schmitt, D. A. (2000). *International programming for Microsoft Windows: Guidelines for software localization with examples in Microsoft Visual C*. Redmond, WA: Microsoft Press.

Sprung, R. C. (Ed.). (2000). *Translating into success: Cutting edge strategies for going multilingual in a global age*. American Translators Association Scholarly Monograph Series XI. Amsterdam: John Benjamins.

Stewart, E. C., & Bennett, M. J. (1991). *American cultural patterns: A cross-cultural perspective*. Yarmouth, ME: International Press.

Symmonds, N. (2002). *Internationalization and localization using Microsoft .NET*. Berkeley, CA: Apress L. P.

Taylor, D. (1993). *Global software: Developing applications for the international market*. New York: Springer-Verlag.

Tuthill, B. (1993). *Solaris: International developer's guide*. Mountain View, CA: Sun Microsystems.

Tuthill, B., & Smallberg, D. (1997). *Creating worldwide software* (2nd ed.). Mountain View, CA: Sun Microsystems.

The Unicode Consortium. (1996). *The Unicode standard, version 2.0*. Reading, MA: Addison-Wesley.

Uren, E., Howard, R., & Perinotti, T. (1993). *Software internationalization and localization: An introduction*. New York: Van Nostrand Reinhold.

Watkins, J. (2000). *The guide to translation and localization: Preparing products for the global marketplace* (3rd ed.). Portland, OR: Lingo Systems.

X/Open Company Limited. (1993, 1994). *Internationalization guide, version 2*. Englewood Cliffs, NJ: Prentice.

Yunker, J. (2002). *Beyond borders: Web globalization strategies*. Indianapolis, IN: New Riders.

ORGANIZATIONS

American Translators Association (ATA): www.atanet.org

Globalization and Localization Association (GALA): www.gala-global.org

Localisation Industry Standards Association (LISA): www.lisa.org

Professional Association for Localization (PAL): www.pal10n.net

The Localization Institute: www.localizationinstitute.com

The Multilingual Internet Names Consortium: www.minc.org

Unicode Consortium: www.unicode.org

W3C Internationalization Activity: www.w3.org/International

International Workshop on Internationalization of Products and Systems (IWIPS): http://www.iwips.org/

MAGAZINES

ATA Chronicle: www.locguide.com/references/publications/magazines

Language International: www.language-international.com

MultiLingual Computing: www.multlingual.com

Translation Journal: www.accurapid.com/journal

Author Index

Subject Index

Printed in the United States
87013LV00003B/4/A